Face De-identification: Safeguarding Identities in the Digital Era

Yunqian Wen • Bo Liu • Li Song • Jingyi Cao •
Rong Xie

Face De-identification: Safeguarding Identities in the Digital Era

Springer

Yunqian Wen
Department of Electronic Engineering
Shanghai Jiao Tong University
Shanghai, China

Li Song
Department of Electronic Engineering
Shanghai Jiao Tong University
Shanghai, China

Rong Xie
Department of Electronic Engineering
Shanghai Jiao Tong University
Shanghai, China

Bo Liu
School of Computer Science, University of Technology Sydney
Ultimo, NSW, Australia

Jingyi Cao
Department of Electronic Engineering
Shanghai Jiao Tong University
Shanghai, China

ISBN 978-3-031-58221-9 ISBN 978-3-031-58222-6 (eBook)
https://doi.org/10.1007/978-3-031-58222-6

This Springer imprint is published by the registered company Springer Nature Switzerland AG
The registered company address is: Gewerbestrasse 11, 6330 Cham, Switzerland

"In recognition of those committed to safeguarding identities and advancing privacy in the digital realm. Your dedication to the ethical use of technology shapes a future where innovation coexists harmoniously with personal privacy."

Preface

Welcome to "Face De-identification: Safeguarding Identities in the Digital Era." As the author/editor of this book, I am honored to present this comprehensive exploration into the intricate realm of safeguarding identities in an increasingly digital landscape.

The idea for this book stemmed from a deep-rooted concern for privacy and security in today's technologically advanced world. The scope of this work encompasses an extensive study of face de-identification techniques, aiming to address the critical challenges faced in protecting identities amid the pervasive use of facial recognition technologies.

Our intent with this book is to offer a thorough examination of various face de-identification methodologies, elucidating their intricacies, strengths, and limitations. Through a structured approach, we have endeavored to present an array of techniques, from obfuscation-based methods to advanced deep generative models, catering to a diverse audience interested in understanding the multifaceted aspects of preserving privacy in digital spaces.

This book is designed for scholars, researchers, practitioners, policymakers, and individuals curious about the intersection of technology and privacy. It serves as a resource for academics delving into the complexities of identity protection, professionals implementing privacy measures, and enthusiasts seeking a deeper understanding of face de-identification in an evolving digital world.

Sydney, NSW, Australia
November, 2023

Bo Liu

Acknowledgments

This work was supported by the Fundamental Research Funds for the Central Universities, the MoE-China Mobile Research Fund Project (MCM20180702) and the National Key R&D Project of China (2019YFB1802701).

About the Book

Our combined team from University of Technology Sydney (UTS) and Shanghai Jiao Tong University (SJTU) started to work on the topic of face de-identification from 2020. Our findings in state-of-the-art de-identification technology have been invaluable, making the insights and perspectives highly commendable and respected.

In this compelling work, the reader is presented with an insightful journey into the world of face de-identification. As the team's engaging narrative unfolds, you will be guided through the intricate landscape of safeguarding identities in the digital era.

As an expert in privacy protection, I have witnessed the evolution and impact of technologies on our daily lives, especially with respect to privacy and security concerns. The exploration of face recognition and de-identification techniques in this book is timely and essential in our increasingly interconnected world.

This book introduces a comprehensive exploration of face de-identification techniques, shedding light on the complexities and challenges faced in this field. The innovative strategies and ethical considerations presented here mark a significant step forward in the ongoing dialogue on privacy and identity protection. I am confident that this work will contribute significantly to the discourse on privacy and technology, fostering deeper insights and inspiring further advancements in this crucial area.

I commend all my co-authors for their dedication and expertise in compiling this significant contribution. It is my privilege to introduce this impactful work to readers and commend its relevance, timeliness, and scholarly merit.

Contents

Acronyms

ϵ	Privacy Budget
2D	Two-Dimensional
3D	Three-Dimensional
3DMM	3D Morphable Model
A^3GAN	Attribute-aware Anonymization Network
AAD	Adaptive Attentional Denormalization
AAM	Active Appearance Model
AE	Auto Encoder
AINet	Attribute-aware Injective Network
AU	Action Unit
BFR	Blind Face Restoration
BIPA	Biometric Information Privacy Act
CA	Coded Aperture
CCPA	California Consumer Privacy Act
cGAN	conditional Generative Adversarial Network
CNN	Convolutional Neural Networks
CS-SFT	Channel-Split Spatial Feature Transform
DGN	Deep Generative Network
DNN	Deep Neural Network
DP	Differential Privacy
DRRDN	Deep Robust Representation Disentanglement Network
FACS	Facial Action Coding System
FATM	Facial Attribute Transfer Model
FID	Fréchet Inception Distance
FIP	Facial Identity-Preserving
GAN	Generative Adversarial Network
GDPR	General Data Protection Regulation
HiSD	Hierarchical Style Disentanglement
ISR	Inverse Super Resolution
JPEG	Joint Photographic Experts Group
LDP	Local Differential Privacy

LPIPS	Learned Perceptual Image Patch Similarity
MAE	Mean Absolute Error
MfM	Multi-factor Modifier
MMDA	Multimodal Discriminant Analysis
NeRF	Neural Radiance Field
OPOM	One Person One Mask
PATE	Private Aggregation of Teacher Ensembles
PCA	Principal Component Analysis
PCC	Pearson's Correlation Coefficient
PDPA	Personal Data Protection Act
PI	Perceptual Indistinguishability
PIPEDA	Personal Information Protection and Electronic Documents Act
PPAS	Privacy-Preserving Attribute Selection
PSNR	Peak Signal-to-Noise Ratio
R^2VAE	Replacing and Restoring Variational Autoencoder
RMSE	Root-Mean-Square Error
SOTA	State-of-the-Art
SSIM	Structural Similarity Index Measure
SVD	Singular Value Decomposition
VAE	Variational Auto-Encoder
WGAN	Wasserstein Generative Adversarial Network

Part I
Introduction

Chapter 1
Introduction

1.1 Background and Motivation

In recent years, the world has borne witness to a rapid surge in artificial intelligence technologies, particularly those rooted in deep learning, alongside the widespread proliferation of face recognition applications. This technological renaissance, however, brings with it a pressing concern—privacy [1–4]. Amidst these groundbreaking advancements, faces stand out as one of the most sensitive forms of biological information, intimately connected to personal identity. The essence of face recognition lies in its biometric authentication, a characteristic that is both unique and irrevocable. Yet, the consequences of this technology extend far beyond mere identity verification. On the one hand, when harnessed for cross-referencing with other databases, it unveils a wealth of an individual's sensitive information. A landmark study by Acquisti et al. [5] underscored how faces can serve as the link connecting diverse databases, revealing trails associated with various personas and ultimately undermining privacy. On the other hand, after confirming the identity of a face through face recognition technologies, advanced visual analysis and understanding tools can infer a large amount of sensitive privacy information from the corresponding visual face. For instance, occupation [6] and health status [7]. This poses a serious threat to the security of personal information.

In light of these growing privacy concerns, the field of face de-identification has emerged as a vital research domain within the realms of security and privacy. Face de-identification, a process that conceals facial features while preserving utility for identity-unrelated applications, has found applications in a wide range of scenarios, from anonymizing faces in media interviews and video surveillance [6] to safeguarding privacy in medical research [7], and beyond [8, 9].

The ubiquity of image acquisition in our daily lives—be it sharing personal images on social media, online learning with cameras, or public safety surveillance—renders the need for enhanced privacy protection all the more critical. Existing privacy safeguards often prove inadequate, allowing third parties to collect

Y. Wen et al., *Face De-identification: Safeguarding Identities in the Digital Era*,
https://doi.org/10.1007/978-3-031-58222-6_1

human facial images without consent for large-scale data analysis or questionable applications.

Prominent social media platforms like Google, Facebook, and Shutterfly have faced scrutiny for compromising the privacy of millions by inadvertently leaking private photos to commercial entities, thus embroiling themselves in biometric privacy disputes. Conversely, the need for extensive public facial image datasets to fuel the development of cutting-edge deep learning models has led to the creation of invaluable resources. Yet, these repositories carry inherent privacy risks, resulting in increased restrictions on data sharing. Notably, datasets such as Microsoft's MS-Celeb-1M, Duke's MTMC, and Stanford's Brainwash were, at various times, withdrawn from public access due to privacy concerns.

The growing spotlight on privacy issues has prompted the enactment of stringent laws and regulations, notably the General Data Protection Regulation (GDPR) [10, 11], which prohibits companies from collecting, sharing, or analyzing user data without informed consent. Within the GDPR framework, privacy information encompasses "personal data related to an identified or identifiable natural person," underscoring the paramount importance of protecting personal identity, particularly in the context of facial image data.

This book, "Face De-identification: Safeguarding Identities in the Digital Era," endeavors to explore the multifaceted landscape of face de-identification. It delves into a wide array of methods and strategies aimed at preserving facial sensitive information, notably identity, while retaining utility for applications unrelated to identity. Through a comprehensive examination of this crucial field, we seek to provide both practitioners and researchers with the knowledge and tools necessary to navigate the intricate intersection of technology, privacy, and identity protection.

1.2 Face Recognition and Face De-identification

From the background and motivation, it can be seen that face de-identification is a benign technology born to stop face recognition from invading personal privacy, and the two are in a state of confrontation with each other. In order to design excellent face de-identification technology, a thorough understanding of face recognition technology is a necessary condition. Therefore, next we will introduce face recognition and face de-identification separately.

1.2.1 Face Recognition

Face recognition is a biometric technology that automatically recognizes people's facial features including statistics and geometric features, which is one of the most important applications of image analysis and understanding. Face recognition tasks can be further divided into binary classification and multiclassification. The binary

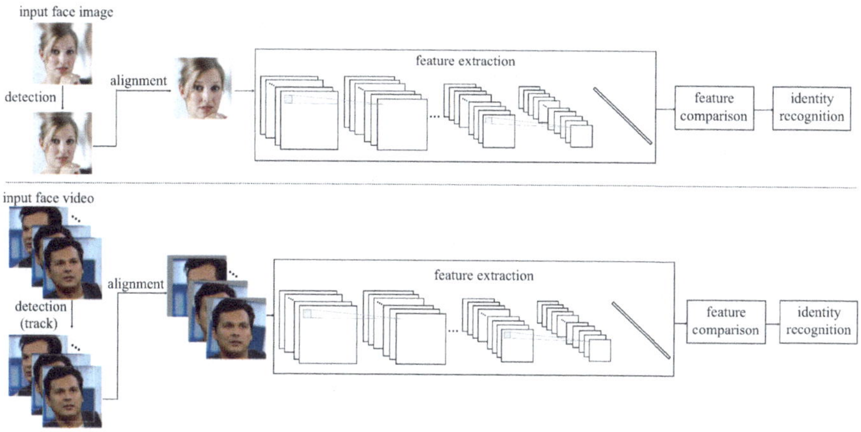

Fig. 1.1 The face recognition process. First, the input image or video is detected and possibly tracks (just for video) to localize the faces. Second, the detected faces are aligned to normalized canonical coordinates. Third, deep facial features are extracted by various methods. After well-designed feature comparison, the identity of the input face data is finally recognized

classification task is also called face verification, which is used to compare whether two images have the same identity. The multiclassification task is also called face retrieval, such as searching for a face with a specific identity in a database of many faces. The widely known face recognition is the abbreviation for identity recognition and verification based on optical facial images. The face recognition process can be simply summarized as using a computer to analyze a face video or image. Firstly, it detects and possibly tracks (just for videos) the faces, so as to localize them. Secondly, it aligns the faces to normalized canonical coordinates. Thirdly, it extracts effective facial features. Finally, it determines the identity of the face object through a comparison of the above-mentioned features. The whole process is shown in Fig. 1.1.

The research on face recognition can be traced back to the late 1960s. The main idea is to design feature extractors and then use machine learning algorithms for classification. Traditional methods rely on hand-made features, such as edge texture description, and combine with machine learning techniques such as principal component analysis, linear discriminant analysis, and support vector machines. The early methods based on geometric features focused on extracting contours and geometric relationships of face components and using the geometric descriptions of shapes and structural relationships as features to construct several feature vectors, including the distance, curvature, and angle between two specified facial keypoints. The advantages are fast recognition speed and low requirements of memory, while the disadvantages are that geometric features can only describe the basic facial information, ignore local subtle features, and result in the loss of local information. The current feature point detection technology is far from meeting the requirements in terms of accuracy.

After introducing deep learning techniques into the field, the approaches have been transferred to extract features with neural networks, which has greatly improved the accuracy and robustness. The deep learning models can be trained by a large amount of data to learn the representation of various variability such as lighting conditions, postures, facial expressions, and so on.

Today, face recognition technology has been widely used in our daily life. Face verification can be treated as a new way of identity confirmation for fast face comparison, mobile payment authentication, security identity verification, etc. Face retrieval can be applied to investigate suspects, complete search of missing persons' databases, and repeated investigation of multiple certificates for one person. At present, the face recognition model can achieve satisfactory accuracy on a specific dataset, but the influence of illumination and posture is still the main challenge. In addition, cross-racial and cross-age recognition problems are also worth studying.

1.2.2 Face De-identification

Due to potential privacy issues, the application of face recognition technology is currently under controversy, and the face privacy protection task is receiving more and more attention. Face de-identification, the main content of this book, is an innovative technical idea to solve the dilemma. There is no consistent definition of de-identification in the existing literature. Ribaric et al. [12] defined de-identification in multimedia content as *"the process of concealing or removing personal identifiers, or replacing them with surrogate personal identifiers in multimedia content."* During this process, other facial features that are not related to identity should remain unchanged, such as expression, posture, and background. After this process, the de-identified face will be judged by the face recognition technology as no longer the same identity as the original face. At the same time, the identity-protected face is expected to retain as much similarity to the original image as possible for normal viewing and sharing and can still be analyzed and processed by other identity-agnostic computer vision methods, such as face detection, motion monitoring, and emotion recognition. Additionally, better image quality and visual effects are also preferred.

With face de-identification technologies, visual service providers can use face visual data to carry out legitimate scientific research, business analysis, security monitoring, social sharing, and other activities; ordinary individuals can enjoy the convenience of visual technology without worrying about their other biometric information due to personal identity associated with the disclosure. It effectively alleviates the concerns about personal privacy and security in today's society. To sum up, providing identity protection for facial visual data is the trend of our time, which has great social significance and practical value.

It is recognized that the main purpose of face de-identification is to conceal the identity information of a face. Images and videos are the two main visual data of human faces, and they are also the focus of face de-identification research.

The face image de-identification algorithm can be viewed as a transformation function δ that maps a given face image X to a de-identified image X', aiming to mislead the face recognition model by reducing recognition accuracy. The process can be formulated as

$$\begin{aligned} \delta(X) &= X' \\ s.t.: \mathrm{ID}\{X\} &\neq \mathrm{ID}\{X'\}. \end{aligned} \tag{1.1}$$

Here the $\mathrm{ID}\{\star\}$ indicates the identity of $\star$ determined by the face recognition model.

The face video de-identification algorithm can also be viewed as a transformation function δ that maps a given face video $V = (v_1, v_2, \ldots, v_n)$ to a de-identified video $V' = (v_1', v_2', \ldots, v_n')$, where the faces in the frames with the same serial number in V and V' will be judged as not the same identity by the face recognition model. The process can be formulated as

$$\begin{aligned} \delta(V) &= V' \\ s.t.: 1 \leq i \leq n, \mathrm{ID}\{v_i\} &\neq \mathrm{ID}\{v_i'\}. \end{aligned} \tag{1.2}$$

In the past few years, researchers have proposed a series of face de-identification methods. The initial traditional methods perform perturbation operation on the face region. Recently, more approaches based on deep learning have been proposed to improve the quality of de-identified results.

It is obvious that face de-identification is a newly emerging research topic. Unfortunately, this topic is very challenging due to the need to meet the needs of multiple parties with conflicting interests, as well as the need to deal with advanced and time-honored face recognition technology. Specifically, current research on the protection of facial visual identity needs to address the following three common technical problems.

The first problem is the difficulty of learning high quality identity representation. In recent years, Deep Generative Network (DGN) has made great achievements in the direction of facial visual synthesis. The prerequisite for using its powerful generation ability to help protect visual identity privacy is to learn and obtain facial identity representation. However, faces contain a wealth of biological characteristics. How to obtain pure (known as *disentangled* in the field of deep learning) facial identity representation is crucial to protect identity while not affecting other information. Facial identity is a unique biological characteristic. Other facial visual features, such as hair color, hairstyle, smile, age, gender, skin color, etc., can be divided into discrete categories based on demographic data and intuitive perception. However, because the visual identity of a human face uniquely corresponds to each live person, it cannot be discretely classified like the other visual attributes mentioned above. At present, researchers can only give descriptive definitions of facial visual identity but cannot carry out mathematical modelling. These make it difficult to learn and obtain disentangled identity representations of faces.

The second problem is the difficulty to disable face recognition, preferably without resorting to other real identities in the process. The current recognition accuracy of face recognition technology on test datasets is close to 100%. The strong recognition ability makes it difficult to protect facial visual identity. Different from the recently popular face swapping topic, face de-identification requires that other identity-agnostic attributes should be kept as unchanged as possible while the identity varies. In other words, the appearance of the generated face must be kept as the original one as much as possible, so the effect of identity protection obtained by simply swapping the face with any other character is bad. In addition, in view of the increasingly stringent laws related to the protection of facial identity in recent years, there is a great legal risk in using face swapping and other methods to directly use real human identities as reference to assist in creating fake identities. Researchers have begun to focus on designing methods to generate virtual fake identities without referring to other real identities. Furthermore, de-identification methods are expected to have additional capabilities such as providing theoretical support, recoverability, and interpretability, all of which are challenging.

The third problem is the difficult tradeoff between privacy and utility. It can be seen from the definition of face de-identification that this task requires protecting visual identity while keeping other biometric characteristics unchanged. In other words, the protected face should have identity privacy and can still be used for tasks unrelated to face recognition. Specifically, it is not difficult to simply hide, remove, or replace the true identity in the face visual data. Simple blurring or color block covering is enough. However, how to make the face with modified identity still practical is tough, which means having quality and visual effects that are comparable to the original data. Furthermore, how to keep other identity-agnostic biometric characteristics unchanged as much as possible is very difficult. Generally speaking, the increase in the effectiveness of identity privacy protection will lead to a decrease in the utility of the de-identified results, which is summarized as the well-known **privacy–utility tradeoff** dilemma in this field [13] and is the focus of all face de-identification research works.

1.3 Book Overview

In order to move the face de-identification research forward, this book presents a comprehensive investigation into face de-identification techniques for privacy protection. On top of a comprehensive overview of the main-stream de-identification methods, we also present our latest research outcomes that can effectively anonymize facial images and videos while preserving data utility for downstream tasks.

The book is organized into three main parts. Part I provides an introduction to the problem. It describes the background and motivation for face de-identification, defines key concepts, and summarizes the threat models and regulations.

Part II delves into various face de-identification techniques. It provides an overview of different categories of methods including obfuscation-based methods, k-same algorithm based methods, adversarial perturbation-based methods, and generative model-based methods.

Then, detailed descriptions of novel techniques [14–27] developed by the authors across image and video modalities are then presented in dedicated chapters. This book proposes a total of four schemes to protect the identity privacy of face images and one scheme to protect the identity privacy of face videos. All five solutions address the common technical challenges described in Sect. 1.2.2. In particular, each technology has individual characteristics. The main content of these solutions is shown in Fig. 1.2.

Chapter 4 introduces a de-identification algorithm centered around facial attribute editing, marking the first methodology presented in this book. This approach combines principles from differential privacy theory and k-anonymity to establish privacy metrics, ensuring protection through the indistinguishability of attributes within dataset images. The overall process comprises three stages: facial attribute prediction, privacy-preserving attribute obfuscation, and the generation of de-identified results. In comparison to previous de-identification algorithms based on attribute editing, this approach additionally takes into account the resemblance between the de-identified image and the original, as well as the controllability of the degree of privacy protection. This flexibility enables adjustments in the tradeoff between privacy and utility.

Chapter 5 presents an identity representation manipulation-based technology, which is the second scheme in this book aimed at achieving de-identification for face images. It combines the deep generative network with the traditional differential privacy theory and proposes a three-stage face image identity protection framework. In the first stage, a DGN is trained to disentangle identity representation in the latent space. In the second stage, the ϵ-IdentityDP mechanism based on local differential privacy theory is devised to protect the identity feature. In the third stage, realistic identity protection face images are reconstructed by the frozen trained DGN. This method can provide theoretically guaranteed protection and an adjustable privacy–utility tradeoff for identities. It also has good generalization ability and low computational overhead.

Chapter 5 exclusively concentrates on the de-identification process, aiming to globally control the level of privacy protection. In order to further enhance controllability for individual images and the diversity of de-identification, Chap. 6 introduces an improved latent space identity editing method. Users can achieve personalized and diverse de-identification results by configuring passwords and privacy levels, corresponding to the direction and degree of identity variation. Additionally, it is worth noting that in certain specific scenarios where the use of original images is preferred, such as in criminal cases, the framework proposed in Chap. 6 also includes recoverability. Under authorized conditions, it can reconstruct the original image based on the de-identification results.

It has been found that when processing face images with different expressions and poses by the technology of Chaps. 5 and 6, inexplicable artifacts often appear

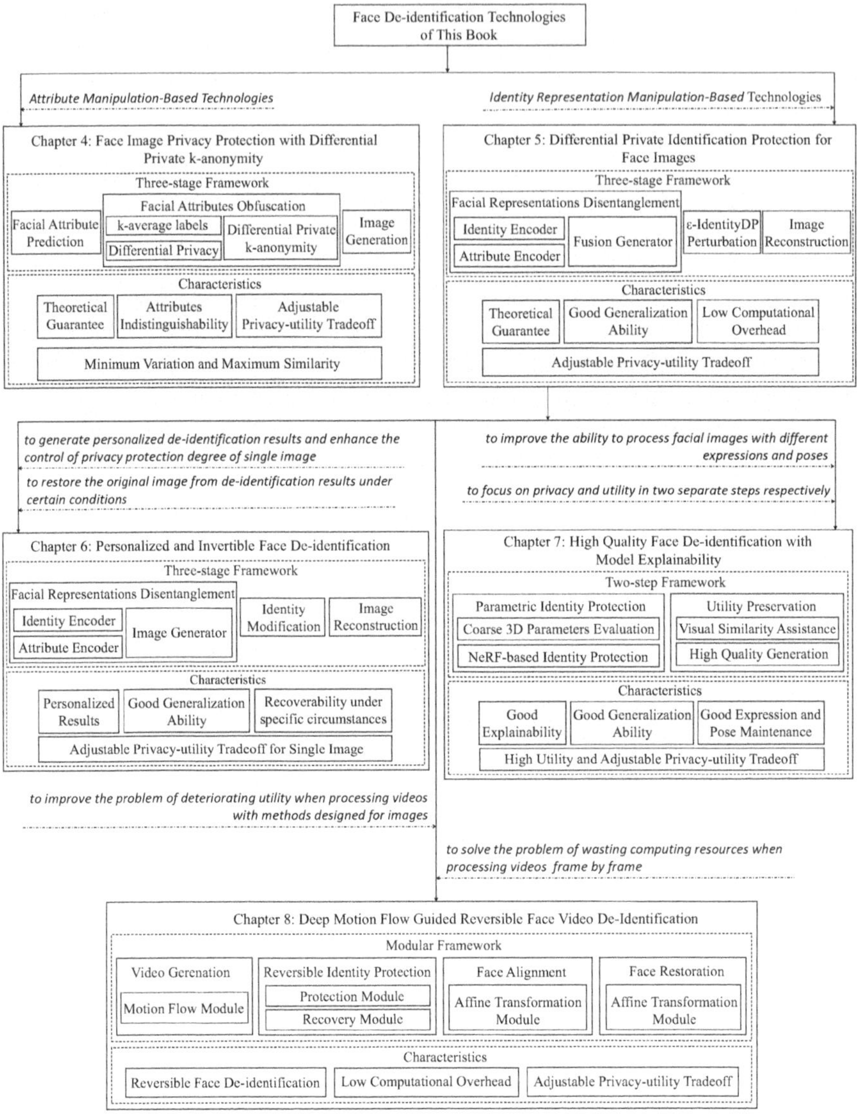

Fig. 1.2 The main content of the five technologies presented in this book from Chaps. 4 to 8

in the generated images. Furthermore, its process of obtaining the identity representation of the latent space through adversarial training in a DGN is cumbersome and lacks interpretability. Therefore, Chap. 7 introduces a two-stage framework with better model explainability, which is the fourth face image de-identification scheme in this book. The first stage is to learn a disentangled identity representation of the three-dimensional (3D) space and protect the identity representation based on a random Gaussian mechanism by an NeRF-based model. The second stage is to obtain high quality realistic de-identified face based on generative priors and parsing maps of the original image. This method uses 3D knowledge and realizes privacy and utility step by step, hence can well maintain various expressions and poses of the original face, and has better explainability. Specially, it can generate results with high utility and provide an adjustable privacy–utility tradeoff, having good generalization ability.

If the method in Chap. 5 is directly applied to face videos, it will be discovered that the utility of the generated videos is seriously deteriorated. In addition, processing face videos frame by frame wastes a large amount of unnecessary computing resources. In order to improve these two problems, Chap. 8 describes a modular framework guided by deep motion flow, which implements reversible de-identification for face videos. The facial motion flow between adjacent frames can be calculated through the *motion flow module*, and then a complete identity-protected (or identity-recovered) video can be produced based on the first frame protected by the *Protection Module* (or recovered by the *Recovery Module*). In order to adapt to the nonstandard poses and expressions, it also designs an effective *Affine Transformation Module* to normalize/restore the first frame face image to the standard/original layout. It is worth noting that the reason why the image processing method in Chap. 7 is not considered here is that this method requires 3D reconstruction of the face to initialize the 3D parameters, so the computational cost is more than that of the DNN-based method. And this is obviously not suitable for video processing. This approach can do reversible face de-identification, has low computational overhead, and can provide an adjustable privacy–utility tradeoff.

Part III concludes the book by discussing future directions and open challenges. It reflects on the progress made as well as opportunities for further advances in this important research field.

In summary, this book makes significant research contributions around designing and evaluating face de-identification models that offer stringent privacy guarantees while retaining utility. The contents offer readers a comprehensive treatment of this problem and provide an organizational structure to easily navigate between introductory material, technical chapters, and conclusions. Researchers and practitioners in multimedia security and privacy will find this a valuable reference.

References

1. B. Liu, M. Ding, S. Shaham, W. Rahayu, F. Farokhi, Z. Lin, When machine learning meets privacy. ACM Comput. Surv. **54**(2), 1–36 (2021)
2. C. Ma, J. Li, K. Wei, B. Liu, M. Ding, L. Yuan, Z. Han, H.V. Poor, Trusted AI in multi-agent systems: An overview of privacy and security for distributed learning. Proc. IEEE **111**, 1097–1132 (2023)
3. G. Zhang, B. Liu, T. Zhu, A. Zhou, W. Zhou, Visual privacy attacks and defenses in deep learning: a survey. Artif. Intell. Rev. **55**, 4347–4401 (2022)
4. B. Liu, M. Ding, T. Zhu, Y. Xiang, W. Zhou, Adversaries or allies? Privacy and deep learning in big data era. Concurr. Comput. Pract. Exper. **31**(19), e5102 (2019)
5. A. Acquisti, R. Gross, F.D. Stutzman, Face recognition and privacy in the age of augmented reality. J. Priv. Confidentia. **6**(2), 1 (2014)
6. A. Senior, Privacy protection in a video surveillance system, in *Protecting Privacy in Video Surveillance* (Springer, Berlin, 2009), pp. 35–47
7. B. Zhu, H. Fang, Y. Sui, L. Li, Deepfakes for medical video de-identification: Privacy protection and diagnostic information preservation, in *Proceedings of the AAAI/ACM Conference on AI, Ethics, and Society* (2020), pp. 414–420
8. J. Lin, Y. Li, G. Yang, FPGAN: face de-identification method with generative adversarial networks for social robots. Neural Netw. **133**, 132–147 (2021)
9. H. Lee, M.U. Kim, Y. Kim, H. Lyu, H.J. Yang, Development of a privacy-preserving UAV system with deep learning-based face anonymization. IEEE Access **9**, 132652–132662 (2021)
10. General Data Protection Regulation (GDPR). Official Text of GDPR. https://eur-lex.europa.eu/legal-content/EN/TXT/?uri=CELEX%3A32016R0679
11. K.A. Houser, W. Gregory Voss, GDPR: the end of Google and Facebook or a new paradigm in data privacy. Richmond J. Law Technol. **25**(1), 1–109 (2018)
12. S. Ribaric, A. Ariyaeeinia, N. Pavesic, De-identification for privacy protection in multimedia content: a survey. Signal Process. Image Commun. **47**, 131–151 (2016)
13. B. Meden, P. Rot, P. Terhörst, N. Damer, A. Kuijper, W.J. Scheirer, A. Ross, P. Peer, V. Štruc, Privacy-enhancing face biometrics: a comprehensive survey. IEEE Trans. Inf. Forens. Secur. **16**, 4147–4183 (2021)
14. J. Cao, B. Liu, Y. Wen, R. Xie, L. Song, Achieving privacy-preserving multi-view consistency with advanced 3d-aware face de-identification, in *Proceedings of ACM Multimedia Asia* (ACM, New York, 2023), pp. 1–6
15. Y. Wen, B. Liu, J. Cao, R. Xie, & L. Song, Divide and conquer: A two-step method for high quality face de-identification with model explainability, in *Proceedings of the IEEE/CVF International Conference on Computer Vision (ICCV)* (2023), pp. 269–272
16. H. Xue, B. Liu, Yuan X., M. Ding, T. Zhu, Face image de-identification by feature space adversarial perturbation. Concurr. Comput. Pract. Exper. **35**, e7554 (2023)
17. Y. Wen, B. Liu, J. Cao, R. Xie, L. Song, Z. Li, IdentityMask: deep motion flow guided reversible face video de-identification. IEEE Trans. Circ. Syst. Video Technol. **32**, 8353–8367 (2022)
18. Y. Wen, B. Liu, M. Ding, R. Xie, L. Song, IdentityDP: differential private identification protection for face images. Neurocomputing **501**, 197–211 (2022)
19. J. Cao, B. Liu, Y. Wen, Y. Zhu, R. Xie, L. Song, Hiding among your neighbors: Face image privacy protection with differential private k-anonymity, in *Proceedings of the IEEE International Symposium on Broadband Multimedia Systems and Broadcasting (BMSB)* (IEEE, Piscataway, 2022), pp. 1–6
20. Y. Wen, B. Liu, R. Xie, J. Cao, L. Song, Deep motion flow aided face video de-identification, in *2021 IEEE International Conference on Visual Communications and Image Processing, VCIP 2021* (2021)
21. J. Cao, B. Liu, Y. Wen, R. Xie, L. Song, Personalized and invertible face de-identification by disentangled identity information manipulation, in *ICCV* (2021)

22. Y. Zhao, B. Liu, T. Zhu, M. Ding, W. Zhou, Private-encoder: enforcing privacy in latent space for human face images, Concurr. Comput. Pract. Exper. **34**, e6548 (2021)
23. Y. Wen, B. Liu, R. Xie, Y. Zhu, J. Cao, L. Song, A hybrid model for natural facede-identiation with adjustable privacy, in *2020 IEEE International Conference on Visual Communications and Image Processing, VCIP 2020* (2020), pp. 269–272
24. J. Yu, H. Xue, B. Liu, Y. Wang, S. Zhu, M. Ding, GAN-based differential private image privacy protection framework for the internet of multimedia things. Sensors **21**(1), 58 (2021)
25. H. Xue, B. Liu, M. Din, L. Song, T. Zhu, Hiding private information in images from AI, in *ICC2020—2020 IEEE International Conference on Communications (ICC)*. Dublin (2020)
26. B. Liu, J. Xiong, Y. Wu, M. Ding, C.M. Wu, Protecting multimedia privacy from both humans and AI, in *2019 IEEE International Symposium on Broadband Multimedia Systems and Broadcasting(BMSB)* (2019)
27. B. Liu, M. Ding, H. Xue, T. Zhu, D. Ye, L. Song, W. Zhou, DP-Image: Differential Privacy for Image Data in Feature Space (2021). arXiv preprint arXiv:2103.07073

Chapter 2
Facial Recognition Technology and the Privacy Risks

2.1 Face Recognition Technology

Facial recognition technology has advanced rapidly in recent years, driven by breakthroughs in deep learning. Deep neural networks (DNNs) now rival and even surpass human performance on facial verification and identification tasks. In this subsection, we provide an overview of key developments in deep-learning-based face recognition algorithms that have propelled progress in this field.

DeepFace [1] was a pioneering deep learning model for face verification and improved face alignment method with an additional 3D model. The pipeline includes detection, alignment, representation, and classification.

DeepID [2] was proposed for multiclassification and to obtain the highly compact and discriminative advanced identity feature by using a small number of hidden variables to represent different identities. In order to obtain effective feature representations to reduce intraclass differences and expand interclass differences, DeepID2 [3, 4] proposed to apply deep convolutional networks and simultaneously use recognition features and verification features as supervision. DeepID3 [5] proposed two DNN architectures constructed from the stacked convolution and inception layers proposed in VGGNet and GoogLeNet.

More recent models like FaceNet [6] and SphereFace [7] have focused on mapping face images into compact Euclidean or angular embeddings where distances directly correspond to face similarity. FaceNet [6] used triplet loss to map face images to Euclidean space, and the distance in this space represents the similarity between facial images. SphereFace [7] converts the softmax loss from Euclidean distance into angular interval and introduces the multiplicative angular margin. CosFace [8] also proved the effectiveness of mapping to hyperspherical space, which proposed the normalization of feature vectors and additive cosine margins. Currently, ArcFace [9] was considered to be the most advanced face recognition model and improved its performance by adding angular spacing to get tighter feature distributions and more pronounced decision boundaries.

Y. Wen et al., *Face De-identification: Safeguarding Identities in the Digital Era*,
https://doi.org/10.1007/978-3-031-58222-6_2

Overall, deep learning has driven rapid progress in face recognition. However, there remain challenges in handling pose, illumination, and age variations. Future work could explore meta-learning approaches to learn more generalized feature extractors that are invariant to these factors. Integrating contextual and semantic information beyond raw face images may also improve recognition abilities.

2.2 Threat Models and Privacy Risks

The increasing capabilities of facial recognition raise important privacy concerns regarding mass surveillance, loss of anonymity, and lack of user consent. Key risks like pervasive tracking, unauthorized biometric data collection, insecure databases, and algorithmic bias must be addressed through comprehensive regulations, audits, and public oversight.

Several specific privacy risks associated with facial recognition technology are:

(1) Surveillance and Tracking: Widespread use of facial recognition for monitoring and tracking people's movements enables pervasive surveillance by both governments and corporations. This infringes on privacy rights and civil liberties, with the potential for constant surveillance chilling free speech, assembly, and individual autonomy. Strict regulations are needed to prevent unchecked use of facial recognition technology for mass surveillance.

(2) Lack of Consent: Facial recognition systems deployed in public spaces often operate without informed consent, exploiting people's biometric data without their permission. This violates privacy expectations and should require opt-in consent for ethical deployment. Scenarios like law enforcement accessing driver's license photos to run facial recognition searches have faced opposition over consent violations.

(3) Biometric Data Leaks: Facial recognition systems require aggregating large biometric datasets, which are prime targets for data breaches and cyberattacks. Centralized databases of facial recognition data could enable widespread identity theft and financial fraud if compromised. Decentralized approaches like on-device processing help mitigate this. Data minimization, encryption, access controls, and audits are also important safeguards.

(4) Misuse of Data: There are risks of collected facial biometric data being exploited for purposes other than intended. For example, a retailer using facial recognition for loss prevention could potentially sell their database to advertising firms or data brokers seeking to profile and target customers. Strict limitations and penalties for unauthorized secondary uses are important.

(5) Bias and Discrimination: Facial recognition systems have exhibited demographic biases, with higher error rates for women, minorities, and younger people. This leads to possibilities of denial of services, profiling, and other discrimination based on inaccurate automated decisions. Ongoing audits for bias mitigation are critical for ethical deployment.

(6) Lack of Anonymity: Facial recognition technology can rapidly link someone's real identity to activities, eliminating anonymity. This could expose people's political views, sexuality, health conditions, and other sensitive details without their consent.
(7) Chilling Effects: The possibility of ubiquitous facial tracking can discourage people from participating in public events and exercising rights like protest and free speech. Just knowing they could be identified and located could deter people from attending political meetings, religious services, protests, or healthcare clinics.

As facial recognition technology becomes more pervasive, comprehensive regulations and safeguards are needed to prevent privacy violations and unethical use:

- Strict audits for bias, especially for use in law enforcement, employment, housing, credit decisions, etc.
- Prohibitions on using facial recognition for illegal discrimination based on protected characteristics like race, gender, age, etc.
- Requirements for openness and transparency about where facial recognition is in use and for what purposes
- Guaranteeing individuals' rights to access, correct, and delete their facial biometric data
- Requiring opt-in consent for facial recognition enrollment and identification, avoiding exploitation of data like driver's license photos
- Assessing whether less intrusive alternatives like badges or keys could meet business needs vs. facial recognition
- Legal protections and penalties for unauthorized access, retention, or misuse of biometric data
- Decentralized approaches using on-device processing and encryption rather than centralized databases vulnerable to breach
- Oversight bodies and ethical review processes for evaluating facial recognition system proposals, akin to Institutional Review Boards for human subjects research

Overall, careful regulation and technical safeguards are essential to prevent abusive uses of facial recognition that could threaten privacy, enable discrimination, and erode civil rights and liberties. A collaborative approach balancing innovation and individual rights will help guide the responsible development of this powerful but potentially dangerous technology.

2.3 Regulations and Acts on Facial Data Privacy

Facial data privacy is an increasingly important area of concern, and regulations related to it can vary by country and region. Here are some of the key regulations and acts that were relevant to facial data privacy:

European Union—General Data Protection Regulation (GDPR) [10] The GDPR, applicable in the European Union, includes provisions related to the processing of biometric data, which includes facial recognition data. It places strict requirements on obtaining consent and ensuring the security and privacy of such data. The GDPR has spurred increased investment in privacy-preserving techniques by EU tech companies. However, ambiguity around consent and legal bases for facial recognition systems in public places remains a challenge.

United States—California Consumer Privacy Act (CCPA) [11] The CCPA, applicable in California, grants consumers rights over their personal information, which includes biometric data. It requires businesses to disclose what data they collect, give consumers the right to opt-out, and provide safeguards for sensitive data.

United States—Illinois Biometric Information Privacy Act (BIPA) [12] BIPA is a state law in Illinois that imposes strict requirements for collecting, storing, and using biometric data, including facial recognition. It has been the basis for several lawsuits against tech companies. Overall, the United States lacks comprehensive protections comparable to GDPR; privacy advocates and states pushing for stronger regulations.

Canada—Canadian Privacy Laws [13] Canada has privacy laws at the federal and provincial levels that govern the collection and use of personal information, which may include biometric data. The federal law, the Personal Information Protection and Electronic Documents Act (PIPEDA), and provincial laws set the standards for data protection.

Australian Privacy Act [14] The Australian Privacy Act governs the handling of personal information, including biometric data, by organizations and government agencies in Australia.

Indian Data Protection Bill (Draft) India was working on a data protection bill that, when passed, is expected to regulate the processing of biometric data, including facial recognition, in the country.

Singapore Personal Data Protection Act (PDPA) [15] The PDPA in Singapore regulates the collection, use, and disclosure of personal data, including biometric data.

Chinese Personal Information Protection Law [16] China was in the process of drafting a comprehensive personal information protection law, which would likely include provisions related to biometric data.

Overall, facial recognition regulation remains uneven globally. The EU has led with GDPR, but other regions are scrambling to catch up. There is a need for international coordination and ethical frameworks given global data flows. A balanced approach that enables innovation while empowering user rights and providing oversight will be important.

2.4 Conclusion and Future Outlook

The rapid evolution of facial recognition technology enabled by deep learning has yielded transformative capabilities for facial analysis and verification. However, the proliferation of this powerful technology has also raised critical privacy, ethical and regulatory concerns given the sensitivity of facial biometric data.

Privacy risks like mass surveillance, lack of consent, and discrimination must be addressed through technical safeguards like encryption and decentralized processing as well as comprehensive regulations. As facial recognition applications continue expanding, sustained public engagement and oversight will be crucial to ensure ethical development and prevent abusive uses.

Looking ahead, striking an optimal balance between innovation and regulation remains challenging but necessary. With collaborative efforts across technology, policy, legal, and ethics spheres, facial recognition could continue advancing safely in sync with societal values and interests. But this requires commitment to data protection, transparency, nondiscrimination, and preserving individual privacy rights.

If developed responsibly, facial recognition technology holds enormous potential to benefit society in areas like security, accessibility, and convenience. Realizing this potential while avoiding potential harms will hinge on acknowledging and proactively addressing the dual promise and risks of this rapidly evolving capability. Maintaining public trust through ethical technology development and use should remain the guiding imperative going forward.

References

1. Y. Taigman, M. Yang, M.A. Ranzato, L. Wolf, Deepface: Closing the gap to human-level performance in face verification, in *Proceedings of the IEEE Conference on Computer Vision and Pattern Recognition* (2014), pp. 1701–1708
2. Y. Sun, X. Wang, X. Tang, Deep learning face representation from predicting 10,000 classes, in *Proceedings of the IEEE Conference on Computer Vision and Pattern Recognition* (2014), pp. 1891–1898
3. Y. Sun, Y. Chen, X. Wang, X. Tang, Deep learning face representation by joint identification-verification, in *Advances in Neural Information Processing Systems*, vol. 27 (2014)
4. Y. Sun, X. Wang, X. Tang, Deeply learned face representations are sparse, selective, and robust, in *Proceedings of the IEEE Conference on Computer Vision and Pattern Recognition* (2015), pp. 2892–2900
5. Y. Sun, D. Liang, X. Wang, X. Tang, Deepid3: Face recognition with very deep neural networks (2015). arXiv preprint arXiv:1502.00873
6. F. Schroff, D. Kalenichenko, J. Philbin, Facenet: A unified embedding for face recognition and clustering, in *Proceedings of the IEEE Conference on Computer Vision and Pattern Recognition* (2015), pp. 815–823
7. W. Liu, Y. Wen, Z. Yu, M. Li, B. Raj, L. Song, Sphereface: Deep hypersphere embedding for face recognition, in *Proceedings of the IEEE Conference on Computer Vision and Pattern Recognition* (2017), pp. 212–220

8. H. Wang, Y. Wang, Z. Zhou, X. Ji, D. Gong, J. Zhou, Z. Li, W. Liu, Cosface: Large margin cosine loss for deep face recognition, in *Proceedings of the IEEE Conference on Computer Vision and Pattern Recognition* (2018), pp. 5265–5274
9. Ji. Deng, J. Guo, N. Xue, S. Zafeiriou, Arcface: Additive angular margin loss for deep face recognition, in *Proceedings of the IEEE Conference on Computer Vision and Pattern Recognition* (2019), pp. 4690–4699
10. General Data Protection Regulation (GDPR). Official Text of GDPR. https://eur-lex.europa.eu/legal-content/EN/TXT/?uri=CELEX%3A32016R0679
11. California Consumer Privacy Act (CCPA). Official CCPA Text. https://leginfo.legislature.ca.gov/faces/codes_displayText.xhtml?division=1.&chapter=22.8.&part=4.&lawCode=CIV
12. Illinois Biometric Information Privacy Act (BIPA). Illinois BIPA Text. https://www.ilga.gov/legislation/publicacts/100/100-0932.htm
13. Canadian Privacy Laws. Personal Information Protection and Electronic Documents Act (PIPEDA). https://laws-lois.justice.gc.ca/eng/acts/P-8.6/
14. Australian Privacy Act. Official Text of Australian Privacy Act. https://www.legislation.gov.au/Details/C2021C00112
15. Singapore Personal Data Protection Act (PDPA). Official Text of PDPA. https://sso.agc.gov.sg/Act/PDPA2012
16. *China's Personal Information Protection Law.* Australian Government Department of Education. https://www.education.gov.au/download/14672/chinas-personal-information-protection-law/30396/chinas-personal-information-protection-law/pdf. Accessed on 1 Nov 2023

Part II
Face De-identification Techniques

Chapter 3
Overview of Face De-identification Techniques

3.1 Face Image De-identification

Face images refer to images with the human face as the main body, which may include hair, neck, and a small part of the upper body. The backgrounds of the images may be a pure color background plate or a complex natural scene. In this section, we will introduce related work on de-identification of face images. According to the technical means adopted to protect identity, we divide the current face image de-identification methods into four categories: obfuscation-based methods, k-Same algorithm based methods, adversarial perturbation-based methods, and deep generative models-based methods. Below, we will introduce them one after another.

3.1.1 Obfuscation-Based Methods

Many face image de-identification methods that are widely used in daily life are based on obfuscation, and there are mainly four types. The first is blur, which refers to replacing each pixel in the sensitive area of the face by the weighted average of the pixels in its neighborhood; or following the approach of Ryoo et al. [1], the facial privacy-sensitive area is first downsampled by a specified multiple and then upsampled back to its original size. The blurred facial area will become smooth and the details will disappear. The second is pixelation, also known as mosaic, which is to divide the detected privacy-sensitive area of the face into a certain range of units (commonly rectangular units) in the two-dimensional (2D) space. Then the pixels in each unit are taken to the average value of the pixels in their areas [2]. The third is mask, which is to cover the detected privacy-sensitive areas of the face with opaque color blocks, of which black rectangular blocks are the most common. The fourth is pixel-level noise, which refers to adding random perturbations to all image pixels in

Y. Wen et al., *Face De-identification: Safeguarding Identities in the Digital Era*,
https://doi.org/10.1007/978-3-031-58222-6_3

detected privacy-sensitive areas. Gaussian noise is the most commonly used random perturbation. These obfuscation-based methods are widely used in daily life because of their simplicity and ease of operation. You et al. [3] pixelated the facial area to protect identity based on the pretrained face detection network YOLO [4].

However, existing studies have shown that the identity protection provided by these technologies is fragile, and the identity information in the de-identified face images is still in danger of being recognized and leaked [5]. McPherson et al. [6] have proven that face images using obfuscation-based methods to protect identity are ineffective when facing with face recognition techniques based on deep learning, that is, the original identity can still be identified with high accuracy. Even worse, de-identification methods based on obfuscation destroy the utility of the image. Firstly, intuitively speaking, the visual effect of identity-protected images will become worse. Secondly, other identity-agnostic computer vision techniques often do not work well on (or cannot process) these de-identified images. Vishwamitra et al. [7] have demonstrated that blur and mask will affect the perceptual score of the image, and the masked images have an even lower perceptual score.

In addition to the traditional obfuscation-based methods described above, researchers are also exploring more effective obfuscation methods for facial privacy-sensitive areas.

Melle et al. [8] design a reversible scrambling technique suitable for face images to protect identities. This technique uses an adaptive codebook to handle privacy-sensitive areas, where the adaptive codebook consists of a set of background patches (image areas without sensitive information) processed by affine transformation. The main idea of this work is to exploit image self-similarity to encode images and combine the encoding scheme with a scrambling procedure to enhance privacy. The authors demonstrate that the tradeoff between privacy protection and utility can be achieved by varying the intensity of the scramble.

Letournel et al. [9] propose variational adaptive filtering on the face area where keypoints have been detected. This method retains the key facial features (i.e., eyes, lips, and their corners) and better maintains the original facial expressions while hiding the true identity. Later, Rafique et al. [10] also propose a method to reconstruct face images using a trained Gaussian–Bernoulli Restricted Boltzmann Machine to generate models that can hide the true identity without changing the expression.

Yuan et al. [11] design a reversible image visual privacy protection framework based on Joint Photographic Experts Group (JPEG) deformation. During the safe JPEG transmorphing, the selected private areas are applied to most types of regional visual obfuscations, such as masking, blurring, pixelation, inpainting, and warping.

Chriskos et al. [12] develop a method to protect identity by hindering face detection. This method introduces artifacts into face images, such as noise and projection. These artifacts render automatic face detection improbable, while the entire image still retains enough information and is recognizable by humans.

Dadkhah et al. [13] investigate the possibility of applying different half-toning algorithms to avoid automatic face detection and recognition. Half-toning is the method of changing a continuous tone of an image into black and white dots in a way that from the particular distance the change cannot be recognized by human eyes. Besides, the converted images are available for human observation. The authors also investigate the privacy-enhancing impact of multiple half-toning techniques, including Floyd–Steinberg dithering for RGB images, and Stucki dither diffusion, Bayern half-toning, and Jarvis half-toning for grayscale images.

On the basis of achieving de-identification through obfuscation, some studies go further and begin to pursue the provision of theoretically guaranteed identity protection. Among them, Fan proposes a pixelation method based on standard differential privacy (DP) theory [14] and a fuzzy method based on DP theory [15]. Both of which hide privacy-sensitive information by adding controlled randomness to the input images, so as to protect individual characteristics and ensure image sharing with strict privacy guarantees. These two methods are shown to effectively reduce the success rate of reidentification attacks.

Later, in order to obtain better image quality, Fan proposes another image confusion solution based on metric privacy [16], a rigorous privacy notion generalized from DP. This method designs a random sampling mechanism that satisfies metric privacy, and uses singular value decomposition (SVD) to generate visually similar images with the same singular matrix but different singular values. Compared with the previous two methods [14, 15], the visual effect of [16] has been significantly improved. However, the perceptual information captured through SVD is limited, and the practicality of the generated images is still not ideal.

Recently, Fan et al. demonstrate an interactive framework for obfuscation of face images in their work [17]. It integrates widely used image quality evaluation methods and practical face recognition technology. Users can view the performance of methods [14, 16] and the other two comparative methods on a dataset of real-world face images. In addition, Liu et al. also propose an identity protection method guaranteed by DP theory by adding global noise [18].

In summary, the obfuscation operations will be reflected at the image level. The operations are not selective but will indiscriminately confuse all facial biometric features in areas that determined to be privacy-sensitive. While the identity is protected, information about other facial visual features is also hidden, and the resulting identity-protected images are often considered to have only limited (or no) utility. To make matters worse, some studies [19, 20] have proven that even face images that are obfuscated through some carefully designed methods are still likely to be identified by the face recognition model again. This makes obfuscation-based methods unreliable. However, from a computational perspective, obfuscation-based methods are mostly simple operations, so techniques from this category are very suitable for use in low-resource situations where the specific image and subsequent use of the face are not important.

3.1.2 k-Same Algorithm Based Methods

In order to improve the comprehensive performance of protecting the identity of faces, methods based on the k-same algorithm are proposed. Before the rise of deep learning, k-Same algorithm based methods once dominated the field of face image de-identification. As strong competitors to the obfuscation-based methods discussed in the previous subsection, the advantage of this type of method lies not only in better utility of the protected images but also in providing theoretically guaranteed identity protection.

The work [21] of Newton et al. first proposes the k-Same algorithm to protect the identity of face image, whose specific steps are as follows. First of all, the dataset of face images is divided into clusters with size k based on the distance metric according to the facial features. Then, each face image to be processed will be replaced by the aggregation of its own cluster, i.e., the average face. Here, Newton et al. design k-Same-Pixel algorithm (averaging the original image pixel by pixel) and k-Same-Eigen algorithm (averaging the projected image of the original image) to do the aggregation. Because all k images in each cluster are represented by the same aggregated face so as to protect the real identity information, the k-Same algorithm gets its name. In addition, since each identity-protected face image appears k times in the entire privacy-preserving image dataset, and it can only match at most one of the k original faces, the k-Same algorithm can theoretically limit the risk of being correctly identified to 1/k. However, due to the small alignment error between the faces in the cluster, artifacts often appear in the images generated by the k-Same algorithm.

Since then, many variants of the k-Same algorithm have been proposed, aiming to improve the utility of identity-protected face images, especially the visual quality. Among them, Driessen et al. [22] design a k-Same-Eigen-like algorithm that can use parameters to adjust the aggregation effect. Gross et al. propose further expansion methods of the k-Same-Select algorithm [23] and the k-Same-M algorithm [24]. The former divides the face dataset into mutually exclusive face groups based on the selected utility function (for example, measuring face similarity based on facial expression). The k-Same algorithm is then used separately in each group of faces. The latter creates a face by averaging the parameters of the Active Appearance Model (AAM) as a proxy for the original face image. The same team later designs a multifactor identity protection framework in [25], which can obtain better utility. Later, Prinosil et al. [26] analyze the implementation issues associated with the k-Same-M algorithm and propose several heuristic methods to make the de-identified face look authentic while determining the expression or gender of the de-identified person.

During this period, there are many k-Same algorithm based methods for de-identification with the help of the AAM model. Among them, Meng et al. design the k-Same-furthest-FET algorithm [27], which recovers the data utility through transferring/cloning the facial expression from the original to the de-identified face. This algorithm is characteristic of not requiring complicated classifiers or high-level semantic information to describe facial expressions. After that, the same team makes many expansions centered on this algorithm and successively proposes the k-Same-furthest algorithm [28] and k-Diff-furthest algorithm [29]. The two derived algorithms pursue a better balance between privacy and utility from multiple perspectives. In addition, by applying the same identity change to all face instances of the same person, the team also proposes a face video identity protection method based on the k-Diff-furthest algorithm [30]. However, due to the interframe changes are not considered, the quality of the de-identified video is very mediocre.

Chi et al. [31] propose an identity subspace decomposition method, which aims to depose the AAM feature space into an identity sensitive subspace and an identity insensitive subspace. After this, the sensitive identity information is separated from the identity-agnostic information, thereby effectively protecting the identity while maintaining the high utility performance of the resulting average face. Sim et al. [32] propose to use a subspace decomposition technique called multimodal discriminant analysis (MMDA) to decouple the AAM parameters of different facial attributes, which in turn enables control of the degree of identity change while keeping attributes such as age, gender, and race constant. In addition, Wang et al. [33] also introduce the work of using MMDA for identity privacy protection. Du et al. [34] propose a GARP-Face method based on AAM model to better preserve the original identity-independent biometric features. After representing the image by the shape and appearance parameters of AAM, Jourabloo et al. [35] select k images that are most similar to the attributes of the test image, formulate an objective function, and finally use gradient descent to learn the optimal weights for fusing k images. The faces aggregated by this method can retain more nonidentity attributes.

With the rapid development of deep neural networks (DNNs), some researchers begin to combine the k-Same algorithm with DNNs and created a series of new methods. They hope that DNN's ability to generate realistic images will better address the challenging privacy–utility tradeoff problem. Among them, Chi et al. [36] first extend their previously proposed subspace decomposition technique [31] to a deep learning model. The new model extracts identity representations known as facial identity preserving (FIP) features from input images and reconstructs faces from the average FIP features calculated based on the k-Same algorithm, resulting in faces with removed real identities that maintain good utility. After that, a series of methods with similar patterns have been proposed successively, such as k-Same-Net [37, 38], k-Dive-Net [39], K-samesiamese-GAN [40], AnonFACES [41], Chuanlu et al. [42], and FICGAN [43]. The utility performance of these methods to generate images has been significantly improved.

In summary, the k-Same algorithm based method is a specific implementation of k-anonymity privacy theory [44, 45] in the context of face data. They use the statistical information of a set of face images to generate more realistic identity-protected face images. Although these methods have good theoretical support and were once the technical pillar in the field of de-identification, they have significant limitations.

Firstly, the k-Same algorithm assumes that each subject appears only once in the dataset, which may not hold true in practice. In real-world scenarios, the presence of multiple images from the same subject or images with similar biometric characteristics may result in lower privacy protection effect than the theoretical level.

Secondly, the k-Same algorithm operates on the closed sets and produces corresponding identity-protected sets, which is not applicable for processing a single image or image sequences.

Thirdly, the identity privacy-preserving results of such methods always look unnatural, let alone maintaining as many nonidentity features as possible with the corresponding original image. The de-identified face image will be significantly different from the original input image and therefore may lose unique features related to race, gender, age, expression, or posture, etc. These situations have been alleviated to a certain extent after the introduction of DNN, but there is still much room for improvement.

Fourthly, the k-Same algorithm is sensitive to composition attacks [46] and attacks using background knowledge [47], so the privacy guarantee provided by this algorithm is not reliable enough in the face of these above two attacks.

Fifthly, the k-Same family algorithm is not suitable for processing face videos. The k-Same algorithm requires that the input image set is specific to an individual, that is, in the image set, each person can only have one image, and there are no two images related to the same person. This makes the k-Same algorithm based methods generally not applicable to a set of frames taken from the same video sequence, since they usually contain facial images of the same person. Even if the method [30] achieves face video de-identification by applying the same identity change to all face instances of the same person, the utility performance of the identity-protected video will be significantly reduced compared to the original video. Therefore, new algorithms must be developed to protect identity privacy in face video sequences. The above five limitations indicate that there is still a long way to go in the research of face image de-identification.

In the past six years or so, research progress on face image de-identification has made a huge leap. A series of new technologies that can better address the privacy–utility tradeoff problem have been proposed, all of which are largely supported by deep learning. The power of deep-learning-based models is that, given a sufficient training dataset, models containing a large number of parameters can be optimized end to end; a well-trained model can strike a good balance between privacy and utility, while being robust to various face recognition technologies. The model parameters can be automatically optimized by mature optimization algorithms through appropriate objective functions designed for input images, output images, intermediate features, etc. With the continuous improvement and development, the performance of this type of model has been significantly improved compared with previous methods. At present, there are two main categories of mainstream new technologies, which are adversarial perturbation-based methods and deep generative model-based methods. We will introduce them one by one in detail below.

3.1.3 Adversarial Perturbation-Based Methods

The development of deep learning has also given new life to face recognition, one of the oldest studied tasks in the field of computer vision. Behind the substantial improvement in recognition accuracy is the development of convolutional neural networks (CNNs) and the availability of large-scale face training dataset [48]. However, the decision-making of CNN models has been shown to be susceptible to interference by adversarial examples, which is produced by adding small perturbations to images [49–51]. As long as the optimization is proper, just disturbing the pixels in the input image with an amount of disturbance that is imperceptible to the human eyes can cause the recognition model to make wrong judgments. The method based on adversarial perturbation can suppress the biometric feature of identity while preserving other nonidentity facial features well. During the training process, this type of model needs to continuously interact with the target face recognition system to weaken its identity recognition accuracy [52]. The final model can mislead the face recognition model by adding small but worst-case disturbances to the face image and produce de-identified faces that are highly similar to the original images.

Some adversarial perturbation-based methods are dedicated to creating adversarial physical devices. Among them, Sharif et al. design a pair of printable eyeglass frame for the target face recognition system [53]. There is about an 80% probability that the identity of the subject wearing this frame can be evaded from being correctly identified or to impersonate another individual. Later, the same research team designs a glass frame based on generative adversarial network (GAN) that can deceive the face recognition system under different imaging conditions (such as different lighting and angle) [54]. Later, Komkov and Petiushko also design a rectangular paper sticker that can be printed on an ordinary color printer [55]. As long as it is pasted on the hat, it can protect the wearer's identity privacy even when facing Arcface [56], one of the state-of-the-art (SOTA) face recognition models. Although these methods do not affect other identity-independent biometric features of the face, they require the user to wear a specific equipment, which is neither convenient nor beautiful, so the application scenarios of these methods are very limited.

More adversarial perturbation-based methods learn to add appropriate perturbations to the image through training. Among them, literature [57] introduces the Penalized Fast Gradient Value Method, which is inspired by the Iterative Fast Gradient Value Method designed for general image [58, 59]. This method operates in the image space domain. The generated de-identified face images will be incorrectly identified with a high probability but can still be viewed and used by users normally, because these images resemble the original ones very much.

During the same period, Oh et al. [60] introduce a general framework based on game theory for the user–recognizer dynamics systems and provide a case study that involves current state-of-the-art adversarial image perturbation methods and person recognition techniques. This method can derive optimal strategy for the user that assures an upper bound on the recognition rate independent of the recognizer's counter measure. Liu et al. [61] combine the idea of adversarial image perturbation that is effective against AI and the obfuscation technique for human adversaries. Deb et al. [62] propose AdvFace, which uses GANs learning to generate minimal perturbations in the salient facial regions. The de-identified faces generated with this method are considered similar to the original images by human observers, and high success rates have been achieved on five advanced automatic face recognition systems. Dong et al. [63] propose an evolutionary attack algorithm that can model the local geometry of the search direction and reduce the dimension of the search space and produce minimal perturbation to the input face image with fewer queries. Zhong et al. [64] proposed DFANet, which applies the dropout layers to the proxy model during each iteration of generating adversarial examples. In this way, as the number of iterations increases, the class integration effect of different generative models can gradually improve the generalization ability of adversarial attacks, which can mislead unseen recognition models. The DFANet can increase the diversity of surrogate models and obtain ensemble-like effects. Zhang et al. [65] propose an adversarial privacy-preserving filter consisting of three modules: an image-specific gradient generator to extract image-specific gradient in the user end with a compressed probe model, an adversarial gradient transmitter to fine-tune the image-specific gradient in the server cloud, and a universal adversarial perturbation enhancer to append image-independent perturbation to derive the final adversarial noise. This filter can add adversarial disturbance to the original image before uploading the photo to sharing services, which is able to mislead malicious face recognition models.

Later, Yang et al. [66] propose a targeted identity protection iterative method, TIP-IM, to generate adversarial identity masks that can be overlaid on facial images to achieve better de-identification performance without sacrificing visual quality. In order to mediate the contradiction between model accuracy and adversarial robustness, the Deep Robust Representation Disentanglement Network (DRRDN) is proposed [67]. This network follows the autoencoder framework and produces robust representations by disentangling from natural adversarial examples. The representation is then aligned to eliminate the effects of adversarial perturbations. DRRDN can obtain adversarial examples with excellent robustness and accuracy. At the same time, the A^3GN method [68] is proposed, which learns instance-level correspondence between faces by adding a conditional Variational Autoencoder (VAE) and attention module to GAN. A^3GN also introduces a face recognition network as a third party to participate in the competition between the generator and the discriminator during training, which allows the attacker to impersonate the target person better. A^3GN can generate natural faces and evade SOTA recognition networks.

Recently, literature [69] proposes a two-stage training method, which firstly uses the attention module to extract the main features of the subject's face and then generates small and almost invisible adversarial perturbations based on the main features of the face, so as to protect the original face image. Zhong et al. [70] believe that it is very unfriendly for a user to generate different protective perturbations for each photo, especially for video frame. So they propose to generate person-specific (class-wise) universal masks by optimizing each training sample in the direction away from the feature subspace of the source identity and name it as one person one mask (OPOM). OPOM generates privacy masks, similar to a customized invisible cloak, by solving an optimization problem that maximizes the distance between diverse deep features of the training images and the feature subspace of the identity. By using OPOM, an average user can only generate one adversarial mask and apply it to all photos and videos of that user, which saves the time of generating adversarial perturbations multiple times for different visual data of the same object.

It is worth noting that although the adversarial perturbations imposed by the above methods can be very small after careful optimization, the original image will generally still have perceptible changes, accompanied by artifacts. In order to make these changes as unobtrusive as possible, some studies combine face de-identification with face makeup tasks, integrating adversarial perturbations into makeup, which allows for a greater perturbation level and achieves identity privacy protection while beautifying the faces. Zhu et al. [71] deceive face recognition models by applying makeup in the eye area. Hu et al. [72] introduce a new regularization module along with a joint training strategy to reconcile the conflicts between the adversarial noises and the cycle consistence loss in makeup transfer, achieving a desirable balance between the attack strength and visual changes. This method hides adversarial perturbations in full-face makeup, improving visual enjoyment while achieving identity protection with a high success rate.

In addition, there are some adversarial perturbation-based methods that make the image "poisonous" by adding elaborate perturbations and then protect the identity privacy of face images by training the targeted recognition model on a dataset containing the poisonous images. The recognition model that has been trained specifically will not be able to correctly determine the identity of a face.

Shafahi et al. [73] adopt a poisoning attack and present an optimization-based method for crafting poisons and show that just one single poison image can control classifier behavior when transfer learning is used. After full end-to-end training under a "watermarking" strategy that makes poisoning reliable using approximately 50 toxic training instances of the selected person, they can control the recognition model's judgment of the selected person's identity during the test without reducing the recognizer's performance in discriminating other objects. Zhu et al. [74] introduce a new "polytope attack" in which poison images are designed to surround the targeted image in feature space, so as to mislead the recognition model. In this attack, the authors generate multiple toxic images from the base class by applying small perturbations, which cause the toxic images to capture the target image within a convex polyhedron in the feature space, i.e., the poisonous images surround the target image in the feature space. After injecting poisonous images into the training dataset, a model trained on this dataset even with unknown architecture and parameters will fail to identify the target face. Similarly, Shan et al. [75] design the Fawkes model to help individuals put on unnoticeable "cloaks" (imperceptible well-designed pixel-level changes) before releasing their photos, so as to combat unauthorized facial recognition models. When these images are used to train a facial recognition model, the trained model will be unable to correctly identify the normal images of these users.

Since the perturbation imposed on the image by the method based on adversarial perturbations is almost visually imperceptible, i.e., the utility of the generated image is very good, and its de-identification ability is always excellent when facing the target face recognition model, that the probability of identity being misjudged is very high, so the adversarial perturbation-based method is currently one of the most popular methods. However, this type of method still has the following key issues that need to be improved:

- The training of adversarial perturbation-based methods often requires various interactions with the target system, so this type of method can only be guaranteed to be effective when it is used for specific and trained face recognition systems, that is, the generalization ability of such models is always weak. In practical applications, it is not known in advance what kind of recognition model will be used to invade privacy, let alone interacting with it multiple times during the training process such as frequent querying. Therefore, the application scenarios of such methods are very limited at the beginning, and this problem has not been alleviated until the researches on model generalization ability are paid enough attention to in recent years.
- When adversarial noise is added to the face image, this type of method requires continuous optimization. As a result, the calculation complexity is generally high and the process is usually time-consuming, which makes them difficult to be applied to large-scale datasets.
- The generated images sometimes still have artifacts, and when the degree of optimization is high, inexplicable weird spots will appear on the image, so the quality of the generated de-identified images still needs to be improved.

The latest research on this type of approach strives to solve the above three aspects. Once tackled effectively, this type of approach can shine. Furthermore, none of the adversarial perturbation-based method has yet emerged to provide theoretically guaranteed protection.

3.1.4 Deep Generative Model-Based Methods

Different from the characteristics of most adversarial perturbation-based methods that they are only effective when facing the specifically trained recognition model, most of the methods in this subsection are also effective for face recognition systems that have never been seen during the training process, that is, they have good generalization ability. Built on deep generative models, these methods can achieve unprecedentedly excellent generation quality and privacy–utility tradeoff driven by large amounts of face image data. In addition, after well-trained, this type of model only needs one forward network processing to protect the identity of face image. So they are less time-consuming and can generate identity-protected face images whose visual quality is comparable to the corresponding original images. Deep generative model-based methods can be further divided into three categories: attribute manipulation-based methods, conditional inpainting-based methods, and identity representation manipulation-based methods. Below we will give a detailed introduction to these three subcategories of methods.

3.1.4.1 Attribute Manipulation-Based Methods

Face attribute manuscript, also known as face attribute editing or face retouching, refers to modifying the value of one or more attributes of the face, such as hair color, hairstyle, skin tone, gender, age, whether to smile, add glasses, hats, etc., while other attributes remain unchanged. The required modifications are generally provided in the form of attribute vectors or driving images, and the entire operation process is usually carried out through GANs. For instance, StarGAN proposed in [76], L2M-GAN proposed in [77], and S2FGAN proposed in [78]. Because there are attributes in all facial attributes that can affect the identity to a greater or lesser extent, changing several facial attributes has the potential to cumulatively modify true identity information. At present, some research uses existing or specially constructed facial attribute manipulation models to achieve de-identification by designing algorithms for manipulating face attributes.

Mosaddegh et al. [79] use direct replacement. They first collect a set of face datasets from donors and then design an optimization strategy to substitute various components (eyes, nose, chin, cheeks, etc.) of the face whose identity needs to be protected with the corresponding facial components of the donors to remove the real identity. In this way, the automatic face recognizer is fooled, while the appearance of the generated face can be as close as possible to the original face.

Li et al. [80] propose the AnonymousNet framework to provide identity protection for images in face datasets. The framework encompasses four stages: facial attribute estimation, privacy-metric-oriented face obfuscation, directed natural identity privacy-preserving face synthesis, and adding adversarial perturbation. Specifically, in the second stage, the authors design a Privacy-Preserving Attribute Selection (PPAS) algorithm to select and update all 40 facial attributes, changing the identity while making the distribution of any attribute close to the true statistical distribution of the attributes in the dataset. AnonymousNet is able to generate natural images with forged identities, and its de-identification result is guaranteed by the t-closeness privacy theory [81].

Pan et al. [82] introduce a reversible face image identity protection framework based on conditional encoder and decoder framework and name it multifactor modifier (MfM). This framework consists of a style encoder, a content encoder, and one decoder. Users only need to select the input image that provides a reference style, set the password, and choose the desired multifactor combination to get the identity-protected image that removes the real identity and meets the user's requirements for attribute. Besides, when the correct password and selected multifactor combination are given, the original face can be restored. It is worth noting that when designing the multifactor attribute requirements, the authors divide facial attributes into identity-related facial attributes (such as gender, age, facial expression, etc.) and identity-independent facial attributes (such as hairstyle and skin color, etc.), which are processed separately.

Zhai et al. [83] believe that the face identity is a high-level semantic representation determined by a combination of specific face attributes, so it can be changed by manipulating face attributes. They propose an attribute-aware anonymization network (A^3GAN), which first uses a multiscale semantic suppression network with a creative suppressive convolution unit to gradually remove face identity along multilevel deep features. Then the attribute-aware injective network (AINet) generates various attributes in a controllable manner and injects them into the latent space code of the original face. Finally, the decoder produces realistic and high quality faces. This method maintains original image quality while protecting identity privacy and allows for more fine-grained control over the de-identification process. However, the result is very unlike the original image.

In summary, attribute manipulation-based methods have relatively mature data with attribute labels for training, and there are also popular face attribute editing methods that can be used for reference, so the research foundation is excellent. The most critical part of designing this type of methods is to design specific algorithms for manipulating attributes, so as to ensure that identity privacy is protected while minimizing the impact on the utility of the image. However, there is a very serious problem with this kind of method, that is, from a common sense perspective, except for age, gender, race, and a few other attributes that affect the whole face and are strongly related to identity, most of the rest of facial attributes are considered to be irrelevant or weakly related to identity, including hairstyle, skin color, whether to smile, eyebrows shade, and so on. And manipulating attributes that are strongly related to identity will greatly affect the utility of the image, especially the similarity with the original face. Therefore, even this type of method ensures that identity is protected, it has a high probability of affecting other nonidentity features of the image, and it is difficult to be as similar as possible to the original image. Last but not the least, changes in facial attributes will also prevent subsequent identity-independent computer vision applications of facial analysis from being used normally.

This book believes that the application scenarios of attribute manipulation-based methods are very limited. For them, identity protection algorithms for attribute editing and retention are customized for specific application scenarios, quantitative analysis of the impact of face attribute editing on face identity, and the introduction of other new means (such as homomorphism encryption) to solve the problem of privacy–utility tradeoff may be the future directions.

3.1.4.2 Conditional Inpainting-Based Methods

A missing/damaged portion/area of an image is a set of unconnected pixels surrounded by a set of known adjacent pixels. Face inpainting refers to the method of using known information to fill or reconstruct the missing/damaged areas of the face image [84]. As for face images, identity information is concentrated in the facial area, this can be proved by the fact that the current face recognition systems require face detection first, thereby framing the face area and then performing subsequent operations only on the detected area. Therefore, if the privacy-sensitive face area (the entire face area including hair and part of the upper body, or demarcated facial area including the five senses and skin by a segmentation algorithm) is removed to ensure that the real identity information is hidden, then the conditional Generative Adversarial Network (cGAN) is designed to inpaint the whole face based on well-designed "conditions," such as the keypoints, edge contours, image statistical indicators of the removed area, etc., so that the completed face has good utility. This is a very persuasive idea to realize the face de-identification.

Sun et al. [85] propose a two-stage face image de-identification method based on conditional inpainting. The first stage is responsible for obtaining 68 facial keypoints of the original face image. If the input image contains the original face region, facial landmarks are detected using the python dlib toolbox, whereas if the original face region is obscured, facial landmarks are generated using a trained landmark generator. In the second stage, the original face image is blacked out or blurred, and the facial landmarks obtained in the first stage are used to inpaint the face using cGAN. The head-inpainted images can mislead machine recognizers while looking realistic.

The reference [86] proposes a DeepPrivacy method. This method first detects the face and performs sparse pose estimation on the face to obtain 7 facial keypoints, including ears, eyes, nose, and shoulders. Then DeepPrivacy normalizes the pixel values of the image area within the detection bounding box to $[-1, 1]$. Finally, this method uses a trained cGAN to generate the final face based on the above sparse pose estimation. DeepPrivacy ensures 100% removal of privacy-sensitive information in the original face because the model does not touch the face area at all, while keeping the face expressions and movements of the original image basically unchanged. However, the appearance of the generated image will change significantly when compared with the original image.

Qiu et al. [87] first remove the identity information of the face image to be protected through four different measures (black rectangle covering human eyes, face cartoonization, adding Laplace noise, and mosaic) and then use a multi-input generative model based on Variational Autoencoder to fuse these de-identified images to reconstruct realistic images and ensure utility. However, the image generated by this method does not resemble the original image.

Kuang et al. [88] propose a face image de-identification method called DeIdGAN, which can synthesize diverse faces by using the desired shapes and styles. The face image is explicitly obfuscated before being input into DeIdGAN. During the generation process, the semantic segmentation map, the front and the background segmentation map are input as conditions; a facial image of another identity is selected to provide the style reference. The face generated by DeIdGAN is very different from the original face and has good identity protection performance.

However, in terms of utility preservation, DeIdGAN can only guarantee semantic similarity with the original face, and the nonidentity attributes often change.

In summary, since conditional inpainting-based methods directly remove the facial area, the trained model can only access limited conditional information from the original face, so this type of method can often do an outstanding job in protecting identity information. However, precisely because of this reason, the original image information that can provide guidance for the inpainting process of the missing area is very limited. The generated identity-protected face usually can only ensure that the inpainted area is seamlessly connected with the adjacent pixels, the overall view looks real, and the information from the original image as the "condition" is as similar as possible to the original image. But there is no guarantee that the remaining nonidentity features are still the same as the original image, let alone the appearance is similar to the original face. Therefore, conditional inpainting-based methods are only suitable for situations where privacy and utility of specified parts are highly required.

3.1.4.3 Identity Representation Manipulation-Based Methods

Protecting the identity while changing the original face as little as possible is the goal constantly pursued by the face de-identification research. However, no matter the attribute manipulation-based methods or the conditional inpainting-based methods, the current effect cannot achieve this goal. In fact, these two types of methods are powerless in preserving original appearance by their own implementation ideas, i.e., manipulating and transforming facial attributes and inpainting faces according to partial facial information. Therefore, even if there are new developments in the future, it will be difficult for them to remove identity information without affecting the nonidentity features of the original face at the same time.

In order to solve this important problem, researchers of face de-identification refer to the research on face recognition and find a novel and effective idea. As we all know, the face recognition system is usually composed of three parts. The first is the face detection and preprocessing, which is responsible for detecting facial regions and performing face alignment. The second is the identity representation learning, which is responsible for extracting discriminative features from aligned face images through trained deep networks. Finally, the similarity scores of the features are calculated, and the face's identity is determined according to a well-designed matching algorithm [89]. It can be found that the identity representation can be used as a representative to determine personal identity, which is the core of face recognition. Therefore, if a similar face identity representation is obtained in the face de-identification task, targeted operations such as obfuscation, hiding, and replacement can be designed according to its detailed representation form, while (or subsequently) focusing on retaining the utility content such as image quality and facial nonidentity features. By designing a wide variety of utility assurance operations, identity protection can be achieved while affecting other facial characteristics as little as possible. This idea theoretically guarantees the feasibility of a good privacy–utility tradeoff. The key to the research lies in how to design identity representation, perform identity representation learning, and train a deep generation network for high quality face reconstruction.

Most studies design deep networks to achieve face feature disentanglement in the latent space, striving to apply protection measures only to identity features, so as not to affect other face attributes. Among them, Meden et al. [90] replace the original face with an artificial surrogate face generated by a small number of identities. After detecting the face area, they first use the deep face recognition model [91] to calculate the feature vector. Then they match it with the fixed face database of M subjects and select the k closest identities ($k \ll M$), which are then fed to the deep generative network (DGN) to synthesize realistic artificial surrogate face with the visual features of the selected k identities. Finally, de-identification can be achieved by blending the surrogate faces into the input frames. In particular, the generation process of the surrogate face is controlled by a small number of appearance-related parameters, such as posture, skin color, gender, expression, etc. By setting these parameters, users can obtain artificial faces with customizable nonidentity features. Chen et al. [92] propose a model, PPRL-VGAN, that combines VAE with GAN. PPRL-VGAN can explicitly separate identity representation from other image representations in the latent space. After a face is input, its identity representation is replaced by the identity code of the target identity, while other image representations especially the expression information of the original image are retained. At last, a de-identified face image is synthesized with the original expression maintained.

Later, Nousi et al. [93] use deep autoencoders to achieve face identity protection. They first design various methods to fine-tune the encoder part of the standard autoencoder and then forward-passed face images to the modified encoder. This encoder can change the identity in the latent space while preserving other attributes. All other information is then passed to the decoder to reconstruct the new face. The generated face changes the original identity but retains other attributes of the original face; however, it is visually different from the original ones. In addition, Guo et al. [94] present an encoder to map the input face image into a vector in the identity feature space. Then they introduce a large-margin model for the synthesized new identities by keeping a safe distance between the generated identity with both the input identity and existing identities. The generation of de-identified face images is finally completed with certain utility through the trained GAN.

It has been found that sometimes there are gaps between the generated facial area and the background, and some studies have designed methods that can harmoniously blend the identity-protected face with the background. Gong et al. [95] propose a face de-identification method that can preserve multiple attributes. They first establish a twofold chained architecture called replacing and restoring variational autoencoders (R^2VAEs) to disentangle identity-related and identity-independent features in the latent space. Then they employ two feature obfuscation strategies to replace the original identity-related feature for synthesizing a de-identified face. At this time, if the identity-protected face is used to directly replace the original face, color differences between the face area and the surrounding area in the image will be caused. Therefore, the authors use de-identified faces as prior information to guide a GAN-based network to fill the original face region, so as to generate de-identified faces with reasonable semantics and realistic features. Kuang et al. [96] design an end-to-end network that relies on the Face Region Loss to achieve identity protection. Specifically, the loss includes a pixel-level loss that constrains the input and output faces to be close to a certain empirical value, a variance loss that stabilizes training, and a perception feature loss that penalizes the original and generated faces if they are close to each other in a certain feature space by using the output of the encoder of the U-Net network [97]. Additionally, the network has losses for background regions and conditional inputs, which together with the Face Region Loss train the network to seamlessly replace faces in input images with synthetic faces. The synthetic face will not be correctly recognized and will look realistic and natural. However, the only drawback is that the de-identified result will have a completely different appearance than the original face.

Later, Maximov et al. propose the SOTA model at that time, CIAGAN [98]. CIAGAN trains a generator network with an encoder–decoder structure in an adversarial manner. It first takes the landmark information of the original face and the image excluding the face area as input and encodes it into a low-dimensional space. Then it represents the randomly selected reference identity in the real face dataset as a one-hot vector and inputs it into the bottleneck layer of the network. In an adversarial game with an identity-guided discriminator in a standard GAN-setting, CIAGAN removes identity information from original faces and bodies while generating high quality images that can be used for identity-independent CV tasks. It can be seen that the identity of the generated image is a composition of both the identity hidden in the input face landmarks and the desired reference identity. The same team later proposes a method to separate the de-identification problem into de-identified image generation and image blending and trains an AnonymizationNet and a HarmonizationNet to implement them separately [99]. For any given input image, the AnonymizationNet randomly selects a control identity parameterized by a one-hot vector and mixes it with the original identity so as to create a new unknown identity. Then the HarmonizationNet blends the generated face in order to naturally fit with the background and overall illumination. Yang et al. [100] use the SimSwap face swapping framework [101] to exchange the identity of the face to be protected with another real face, while not changing other biological attributes of the face to be protected. The resulting face is improved by super-resolution reconstruction and generates the high-definition mask face. The authors then design two modules, the Putting on Mask Face architecture and the Putting off Mask Face architecture, that can protect the original face and losslessly recover the original face by authorized users.

In particular, some researchers will borrow existing high accuracy face recognition models and use their output as face identity representation. They are thus able to not worry about designing the data form of identity representation. The research focus is on training DGN to achieve latent space disentanglement and the generation of high quality images.

Wu et al. [102] propose the PP-GAN, a model that inputs the original image and the generated image into the face recognition network FaceNet and extracts the output features after a certain embedding layer as the identity representative of the two images. PP-GAN then calculates the $\mathcal{L}_2$ distance between the two identity representatives as contrastive loss to help remove the real identity information of the original face and create a new identity, so as to achieve the purpose of de-identification. In addition, they design a regulator to maintain image utility as measured by the Structural Similarity Index Measure (SSIM) loss. By designing two types of loss functions that are responsible for identity protection and utility maintenance of the original image, the network can better balance the privacy and utility.

This design pattern inspires a number of subsequent studies. For example, Lin et al. [103] present identity content loss, adversarial loss, and pixel loss, which provide an easier-to-train and more stable face de-identification model for social robot platforms; Khojasteh et al. [104] design de-identification loss and perceptual loss and use the StyleGAN model [105] as a generation backbone to produce high quality de-identified images without any noticeable distortion; Zhao et al. [106] propose identity-level loss, pixel-level loss, and perceptual-level loss. The weighted sum loss function is used to train a generator based on the StyleGAN2 model [107], which is very effective against face recognition systems.

Furthermore, the network proposed by Cho et al. [108] adopts a VAE with encoder–decoder structure to learn an organized latent space. They use a trained face recognition model [109] to generate guided identity representation, which enables the feature vector produced by the encoder to be divided into identity-related parts and attributes-related parts. This disentangled latent space then allows the identity information to be modified solely of other attributes, and finally, the decoder can effectively generate a natural face with completely new identity, while the other attributes that are loosely related to personal identity are preserved. However, this framework produces images of average quality. Luo et al. [110] also design a VAE-based projector that encodes the identity priors from the face recognition model CurricularFace [111] as a latent variable. Afterward, the authors define the projected identity as the output style and design a carefully devised Adaptive Attribute Extractor to represent the extracted identity-irrelevant attributes as noise input and used the StyleGAN2 model as the generator to improve the quality of face de-identification on megapixels. Cheng et al. [112] train four models that are GoogleNet [113], ResNet50 [109], VGG16 [114], and DenseNet121 [115] to conduct identity, facial expression, gender, and ethnicity, respectively. Then, both Autoencoder (AE) and GANs approaches can be used to change the identity while keeping the other three attributes unchanged, which provides identity privacy protection for 2D and 3D face images.

Some works have achieved reversible face de-identification. Gu et al. [116] obtain a reversible face identity transformer using discrete passwords by optimizing a multitask learning objective function. They define the feature of the face image extracted by a pretrained face recognition model SphereFace [117] as the identity of this face image. During the optimization process, the authors maximize the feature-level dissimilarity between pairs of de-identified faces that have different passwords, and fooling a face classifier, they also maximize the feature dissimilarity between a de-identified face and its identity-recovered face with a wrong password so that the identity always changes. The identity transformer can set a password to protect identity and restore the original face only when the correct password is given.

In addition, some researchers believe that the latent space feature disentanglement cannot be well explained at present. At the same time, the way to model the shape and texture of face images in 3D monocular face reconstruction research is both clear and well explainable. Therefore, it is a reasonable idea to first adopt 3D priors to model face identity parameters and then use the powerful generation capabilities of DGN to generate photo-realistic output. Sun et al. [118] propose a hybrid model, including a face parametric model and a data-driven GAN model. They first use a 3DMM model to perform parametric modelling of the face. After obtaining the semantic parameters of all training set images, they cluster all identity parameters into 15 different identity clusters. Then the distance between the input face and the identity parameters of these 15 clusters will be calculated, and the identity parameters of the input face will be explicitly replaced by the identity parameters of the training set cluster closest to the specified distance. At last, the authors use a GAN to add fine-grained details to the identity-protected rendered faces, improving the overall realism and seamlessly integrating it with the background.

In summary, identity representation manipulation-based method is the best method in terms of comprehensive performance in the current research on face image de-identification and is also the mainstream method based on DGN. The main reason is that this kind of methods possesses the following two advantages:

- Whether it is using DNN to learn face identity representation, or relying on parametric face modelling to obtain face identity representation, as long as the disentangled face representation is obtained, the researchers can only investigate how to protect this representation. Each research scheme shows its own remarkable processing methods. It can directly perform obfuscation operation on the identity representation, or it can design diverse loss functions to make the identity representation far away from the original one. It can be seen that this kind of face de-identification methods is accurate and efficient.
- The excellent performance of DGN in generating high quality face images shows that it is possible to use DGN to turn protected facial identity representation into photo-realistic images with outstanding utility performance.

These two advantages also point out the three research focuses of identity representation manipulation-based methods in the future. To be specific, that is, how to design and obtain a thoroughly disentangled face identity representation, how to protect the face identity representation, and how to design as well as train a DGN with satisfactory image generation quality. In addition, it will become a future trend to give this type of methods more additional functions, such as not relying on real identity assistance during the identity manipulation process, a reversible de-identification process under certain given conditions [142], the

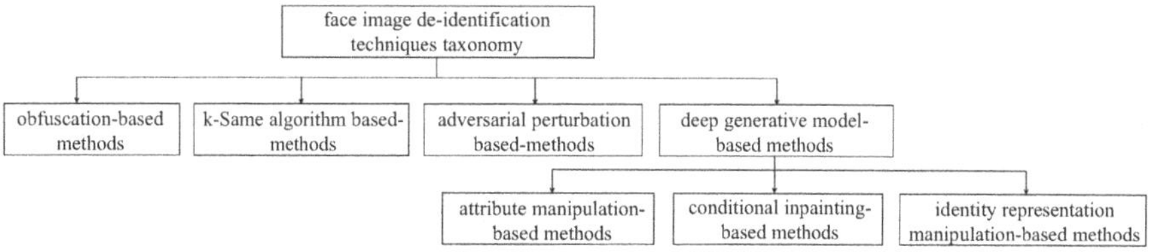

Fig. 3.1 Taxonomy of existing face image de-identification techniques. The proposed taxonomy partitions face de-identification technologies into obfuscation-based methods, k-Same algorithm based methods, adversarial perturbation-based methods, and deep generative model-based methods. The last type can be further divided into three categories: attribute manipulation-based methods, conditional inpainting-based methods, and identity representation manipulation-based methods

diversity of generated results, the adjustability of the degree of de-identification, the interpretability as well as the theoretical support of the proposed method, etc. Researchers are working hard to provide excellent, rigorous, and comprehensive solutions for identity privacy protection.

In particular, since DGN has the ability to encode face images into features in the latent space, it becomes possible to combine DGN with the classic DP theory designed for traditional databases. This has been a promising research direction, and there are currently some methods [119–123] that provide privacy protection guaranteed by DP theory for face images. These methods benefit from the generation ability of DGN, which often results in striking visual effects. Our algorithm classification summary of current face image identity privacy protection research can be seen in Fig. 3.1.

3.2 Face Video De-identification

Different from surveillance videos that contain a large amount of scene information, face videos refer to videos shot with human faces or heads as the main subject, such as vlogs, live streaming sales, speeches, and interviews. The face video may contain a small part of the upper body, and the background may be a pure color background plate or a complex natural scene. In recent years, face video has become very popular on the Internet media, and the number of its creations has increased rapidly. Therefore, the corresponding research on its identity protection is also flourishing. These studies can be divided into two categories, i.e., methods of applying image de-identification methods to videos and methods designed specifically for videos, based on whether they are designed and trained for face videos alone. We will introduce them below separately.

3.2.1 *Methods of Applying Image De-identification Methods to Videos*

For the methods that directly apply face image de-identification methods to videos, we further classify them into two categories: methods of applying image method frame by frame and methods of adding smooth transition measures between frames.

3.2.1.1 Methods of Applying Image Method Frame by Frame

Previously, a considerable number of researchers regarded video as a combination of multiple frames of images. Hence after designing de-identification method for face images, they will state in the corresponding thesis that this method can also be directly applied to face videos. More specifically, they process each video frame as a separate face image, such as the literature [3, 8, 9, 12, 14, 30, 70, 79, 87, 92, 108]. However, this processing idea ignores the unique data characteristics of video. It is worth noticing that there is a close temporal correlation between adjacent frames of face video, and the faces in each frame usually have much more abundant postures and expressions than the general image dataset. As a result, de-identified videos generated through the methods of applying image method frame by frame tend to be of very poor quality.

3.2.1.2 Methods of Adding Smooth Transition Measures Between Frames

Some other researchers also agree with the previous point of view, but they add some measures to facilitate smooth transitions between frames when transferring face de-identification methods designed for images to videos. For example, the method based on identity representation manipulation by Meden et al. [90] is also applicable to face videos. They first process each video frame as an image (see Sect. 3.1.4.3 for details) and then complete two postprocessing tasks. The first is to detect facial landmarks in the generated face and the original input face frame and then use both sets of landmarks to estimate a perspective transformation that warp the artificially generated face to align with the original face. The second is to perform simple skin color segmentation by using the upper and lower boundaries in the HSV color space that define the skin intensities, ensuring that most of the background around the generated facial areas is removed, while only remaining facial areas without the gray-colored background. Finally, the authors blend the warped and segmented synthetic face image with the original image by performing a Gaussian kernel mask.

Additionally, the CIAGAN model [98] proposed by Maximov et al. will first process each video frame as a separate image (see Sect. 3.1.4.3 for details). Then a spline interpolation is used to smooth the face landmarks of each frame between neighboring frames. Finally the temporal consistency of the generated video frames is ensured through a SOTA video translation model with low computational cost.

Another face image de-identification method [99] later proposed by the same team also uses a similar processing approach. After processing each video frame as an image (see Sect. 3.1.4.3 for details), in order to improve the temporal consistency of the video sequence, the authors transform the HarmonizationNet into a frame recurrent network by concatenating the output of the previous frame to the input of the current frame and replacing a spatial discriminator with a temporal one. When processing a video, the temporal discriminator [124] takes three consecutive frames as input and judges temporal smoothness and visual quality. Such simple changes to the HarmonizationNet are experimentally proven to be effective in reducing color jitter in the final results.

In summary, when transferring face de-identification methods designed for images to video, this type of methods adopts a more thoughtful approach by adding operations that are conducive to maintaining video temporal consistency than directly treating the video as a collection of multiframe separate images. The visual effect of the resulting video will be better, without excessive shaking and flickering. However, in fact, the face image domain and the face video domain are collections of two different forms of data, and there is a certain domain gap between them. Therefore, the effect of the above methods (no matter the methods of applying image method frame by frame or the methods of adding smooth transition between frames) is always not ideal. Researchers should design special de-identification solutions for face videos.

3.2.2 *Methods Designed Specifically for Videos*

At present, most of the de-identification methods designed specifically for face video are based on manipulating identity representation assisted by DNNs, and a few are based on obfuscation.

3.2.2.1 Methods Based on Manipulating Identity Representation

This type of methods is based on the same understanding that is obtained through the investigation of deep face recognition research, that is, DNNs can extract disentangled face identity representation in a certain designed latent space, as the face image de-identification methods based on identity representation manipulation in Sect. 3.1.4.3. The video de-identification methods based on manipulating identity representation aim to utilize DNNs to obtain face identity representation and then design targeted obfuscation, hiding, replacement, and other operations according to the representation form. At the same time (or subsequently), they design operations that can maintain utility. Depending on whether the assistance of other real identities is required when generating a new identity, these methods can be further divided into methods based on real identity assistance and methods based on original identity modification.

(1) Methods based on real identity assistance
The method based on real identity assistance refers to replacing the identity of the input face video with a real personal identity of an authorized donor. This type of methods is simple and effective. The way of replacement here can be a direct exchange of two identity representations, or it can be a hybrid of two identity representations into a new mixed identity through various means. Generally speaking, the latter is a more thorough way of protecting facial identity privacy.
Zhu et al. [125] directly use the face swapping technique, Faceswap.[1] They conduct targeted training for Faceswap on the medical video dataset of patients with Parkinson's disease, providing identity protection for patients while liberating the sharing of clinical video materials. Specifically, during the training process, the authors explicitly exchange the face in the open-source dataset with the patient's face so as to protect the patient's identity, while keeping the keypoints unchanged to ensure that the subsequent disease assessment (the face of the patient is a clinical manifestation of movement disorders) and medical analysis can still be conducted according to the de-identified face. However, such a direct face swapping operation will lead to an extreme deterioration of visual similarity, and most nonidentity attributes except for facial keypoints of the result will be different from the original input. Therefore, a series of methods that can better preserve nonidentity features in the de-identification process have been proposed.
Samarzija et al. [126] first train several AAMs to capture and synthesize specific face poses. Then by fitting each model to the input face, they select the best fitting model and determine the face poses and face regions. Finally, the face region is swapped with another face that is extracted from the training dataset used to build the chosen best fitting model. Since this identity exchange takes poses into account, it makes the generated de-identified videos more natural.
Li et al. [127] treat the nonidentity facial attributes as the style of the original face, and they use a trained DNN model, Facial Attribute Transfer Model (FATM), to map nonidentity-related facial attributes to the face of donors. The donors here are several (usually 2–3) real humans who agree to authorize the identity. Using real faces to receive identity-independent features of the face to be protected can ensure that the synthetic face is realistic and natural. Besides, FTAM blends the donors' facial attributes to those of the original faces to diversify the appearance of the synthesized faces.
Gafni et al. [128] propose a feed-forward encoder–decoder network architecture that concatenate the activations of the face-classifier representation layer [109] to the latent space of the network's bottleneck. During the training process, the authors design a number of loss functions, including a new attractor–repeller perceptual term, to complete identity-distancing while maintaining pixel-space similarity. Meanwhile, a multilevel face descriptor is used to describe identity

[1] https://github.com/deepfakes/faceswap

(high-level) and nonidentity features (low or mid-level), respectively. The encoder–decoder network outputs an image and a mask simultaneously, with an extracted transformation matrix by using an estimated similarity transformation (scale, rotation, and translation) to an averaged face. The final output frames are generated by linearly mixing the input video frame and the output image per pixel according to the weight of the transformed mask.

(2) Methods based on original identity modification

Although methods based on real identity assistance have developed to a stage where the protection effect of identity privacy is quite amazing, the generation process requires other real identity assistance, and the new identity in the generated video is (or is mixed with) the identity of one or more existing authorizers. Both of which will make it difficult to apply these methods under increasingly stringent laws and regulations. For example, GDPR stipulates that application providers must periodically obtain the consent of objects who have authorized their identity so as to continue using their identities. So once the authorizer changes his or her mind, the network may need to be fine-tuned, and the previous identity-protected facial video may face the crisis of no longer being used, which is very inconvenient. Therefore, the pattern of extracting disentangled facial identity representation and designing means to make it changed by training a DNN becomes increasingly welcome. We call the method that adopts this pattern "the methods based on original identity modification." Once the identity representation is disentangled from other face attributes, researchers can take steps to eliminate, reduce, or obfuscate the identity representation until the original face identity changes. Simultaneously (or subsequently), researchers can generate realistic face videos based on the new identity representation. During this process, a new virtual identity is born, and no other real identities are required to participate.

Gross et al. [129] describe a framework using multifactor models that unify linear, bilinear, and quadratic data models. They first use the generative multifactor model to factorize the input image into identity and nonidentity components. Then a de-identification algorithm is applied on the combined factorized data. Finally, the bases of the multifactor model are used to reconstruct de-identified images. Experiments on medical record videos after shoulder surgery demonstrate that this approach can protect identity privacy while preserving many of the data utility.

With the rapid development of deep learning, the superb representation learning ability of DNN makes it unique in the face representation disentanglement task. Besides, the ability of DGN to generate high quality images has also aroused the interest of researchers. Ren et al. [130] train two competing systems in an adversarial training setting, that is, a video anonymizer that modifies the original video to remove identity information while still trying to maximize spatial action detection performance, and a discriminator that tries to extract identity information from the de-identified videos. The video anonymizer is trained to learn to disentangle identity representation and action representation. After adversarial training, it can modify the facial identity of the original video without affecting facial movements. Camera systems using this algorithm by designing an embedded chipset at the hardware level can still recognize important events and assist human daily lives by understanding its videos but do not invade personal privacy.

In summary, the methods based on manipulating identity representation are the best comprehensive methods in the current researches on face video de-identification. The corresponding reasons and research key are the same as the identity representation manipulation-based de-identification methods for face images (see Sect. 3.1.4.3 for details). In particular, in addition to the same three future research emphases (see Sect. 3.1.4.3 for details), this type of methods has to focus on two additional research keypoints: (1) algorithm design to adapt to rich expressions and postures and (2) algorithm design to save computational overhead.

3.2.2.2 Methods Based on Obfuscation

There are also a few methods that use obfuscation processing to de-identify face videos. This type of methods generally first employs algorithms to detect privacy-sensitive areas in face videos. Then the detected areas will be confused either by designing an obfuscation algorithm or by using traditional methods such as blurring, pixelation, and color blocking.

Xuan et al. [131] use large-capacity color watermark technology to embed watermark information containing facial privacy area features into the original image. Then they encrypt and decrypt the face area to update the image, thereby achieving efficient identity protection. Korshunov et al. [132] objectively assess five de-identification methods based on obfuscation in detail, namely blurring, pixelation, masking, warping [133], and morphing [134]. When different intensity parameters are selected, the authors investigate the impact on three face recognition algorithms in OpenCV.

In summary, this type of methods protects identity privacy while seriously damaging the data utility of the video; however, the calculation is simple. Therefore, it is suitable for situations where real-time playback is required, or platform resources are limited (such as embedded devices). Our algorithm classification summary of current face video de-identification research can be seen in Fig. 3.2.

3.3 Evaluation Metrics

In this section, we will introduce the evaluation indicators used to measure the performance of the face de-identification algorithm. Unfortunately, there are not yet universally acknowledged evaluation criteria at present, so we sort out the evaluation indicators used in existing papers and present the definition or explanation of the metrics. We divide common evaluation metrics into two categories: the metrics that

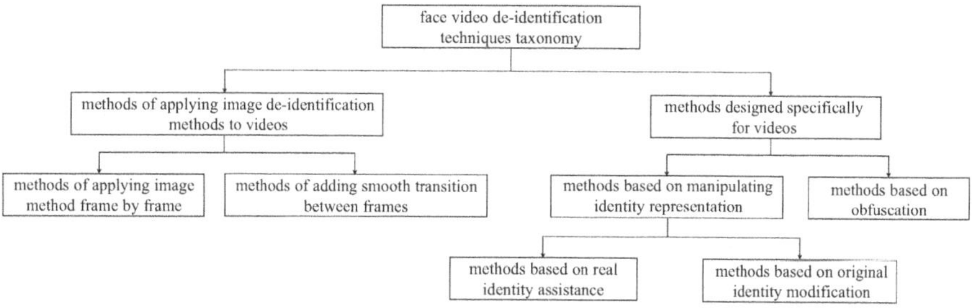

Fig. 3.2 Taxonomy of existing face video de-identification techniques. The proposed taxonomy partitions face de-identification technologies into methods of applying image de-identification methods to videos and methods designed specifically for videos. The former can be further divided into two categories: methods of applying image method frame by frame and methods of adding smooth transition between frames, while the latter can also be further divided into two categories: methods based on manipulating identity representation and methods based on obfuscation. The methods based on manipulating identity representation are subdivided into two classes: methods based on real identity assistance and methods based on original identity modification

evaluate the privacy performance of the algorithm and the metrics that evaluate the utility performance of the algorithm. We will introduce them respectively below. It is worth noting that these indicators are calculated on face images. When evaluating face videos, we calculate the de-identified video against the original video frame by frame.

3.3.1 Privacy Protection

- *Identity Distance*: Almost all face verification models judge whether two images have the same identity by comparing identity embedding distance, so we use the distance between identity vectors e_{id} extracted from the face recognition model, which can be formulated as

$$Id\text{-}dis = Dis(e_{id}(X), e_{id}(\mathcal{F}(X))). \tag{3.1}$$

  Here Id-dis denotes the identity distance, X indicates the original image, and $\mathcal{F}(X)$ represents the de-identification result. The specific form of Dis is determined by the face recognition model used to obtain identity embedding e_{id}, and $\mathcal{L}_2$ distance and cosine similarity are two general measurement functions for calculating the distance.
- *Successful De-identification Rate*: In face verification, when identity distance exceeds the reference threshold given by the model, it is considered that the identities of two images are different. If the identity of the de-identification result is different from the original, it is defined to be a successful de-identification. Therefore, we compare identity distance with the corresponding threshold to fur-

ther calculate the ratio of successful de-identification, which can be formulated as

$$SDR = 1 - \frac{1}{N}\sum_{i=1}^{N} f_{ver}(X, \mathcal{F}(X)), \tag{3.2}$$

where SDR indicates the successful de-identification rate, $f_{ver} = 0$ when *Id-dis*$> \tau$, otherwise $f_{ver} = 1$, and N is the number of testing.

3.3.2 *Utility Preservation*

3.3.2.1 General Utility

- *Face detectability*: The most important point for data utility is that the generated images should look natural and realistic. This can be quantified by evaluating to what extent de-identified faces could be detected by face detectors. Two options are possible to achieve this target: (1) face detection rate (%), i.e., the proportion of de-identified faces that can be detected by a face detector; (2) face detectors like [135] that provide a confidence score for detecting a face, i.e., face detectability could be directly evaluated by comparing the face detection scores obtained for the raw and de-identified elements; alternatively, we could also compare softmax probabilities that the face recognition network produces.
- *Landmark distance*: The localization of face regions and keypoints is a common processing in face modification and other operations, so it is desirable that the deviation of landmarks at pixel level will not be affected in the process of face de-identification.
- *Peak signal-to-noise ratio (PSNR)*:

$$PSNR = 10\log_{10}\left(\frac{MAX^2}{MSE}\right), \tag{3.3}$$

where MAX is the maximum possible pixel value of the image and MSE means the mean squared error between two images I and K with the resolution of $m \times n$, calculated by $MSE = \frac{1}{mn}\sum_{i=0}^{m-1}\sum_{j=0}^{n-1}[I(i,j) - K(i,j)]^2$.
- *Structural similarity (SSIM)*: We want to maintain similarity between the original image and the de-identified image rather than just replace the face area randomly. In other words, we only want to remove the privacy-related characteristics, but keep the visual similarity, i.e., contours and luminous condition. The structural similarity (SSIM) is defined as follows:

$$SSIM = l(X, D(X))^{\alpha} \cdot c(X, D(X))^{\beta} \cdot s(X, D(X)))^{\sigma}, \tag{3.4}$$

where $l(X, D(X))$, $c(X, D(X))$, and $s(X, D(X))$ denote, respectively, brightness similarity, contrast similarity, and structural similarity, and α, β, γ their weighting coefficients, usually set to 1. It should be noticed that both PSNR and SSIM just focus on objective image quality evaluation and fail to capture many nuances of human perception [136].

- *Fréchet Inception Distance (FID)* [137]: FID is used to measure the distance between the distribution of real images and synthesized images. It is a metric that compares the visual quality of generated samples to real ones. The lower the FID, the better, corresponding to more similar real and generated samples. Wang et al. [138] proposed a variant of FID for video evaluation, which measures both visual quality and temporal consistency. Specifically, they infer a video recognition network after removing its last few layers and consider this feature extractor as the "inception" network. For each video, a spatio-temporal feature map is obtained by feature extractor, and then means and covariance matrices are computed for the feature vectors from real and synthesized videos.
- *Learned Perceptual Image Patch Similarity (LPIPS)* [136]: The LPIPS Distance, also known as *perceptual loss*, is used to measure the similarity between two images based on deep features. This metric was proposed to learn the inverse mapping of generated images to ground truth and prioritize the perceptual similarity between them. It has been demonstrated to correlate better with human perceptual similarity than traditional metrics (such as MSE/PSNR, SSIM, FSIM). The lower the value of LPIPS, the more similar the two images are, and vice versa, the greater the difference.

3.3.2.2 Customized Utility

- *Attribute preservation*: To examine the performance of the proposed model for attribute preservation, we depend on separate attribute classifiers for each attribute to calculate the accuracy rate (%) of the preservation performance on some demographic attributes, like hair, gender, age, and skin.
- *Pose preservation*: After de-identification processing, whether the pose of the resulting face is consistent with the original face is an important utility performance, which is related to whether the nonidentity pose-related computer vision tasks can be used normally. Input the de-identified face X' and the original face X into the same third-party pretrained face pose encoder E_{pose}, and the distance between the two output pose feature vectors is used to measure the ability of pose retention. The specific calculation method of the distance here is determined by E_{pose} used.
- *Expression preservation*: After de-identification processing, whether the expression of the resulting face is consistent with the original face is also an important utility performance, which is related to whether the nonidentity expression-related computer vision tasks can be used normally. Input the de-identified face X' and the original face X into the same third-party pretrained face expression encoder E_{exp}, and the distance between the two output expression feature vectors

is used to measure the ability of expression retention. The specific calculation method of the distance here is determined by E_{exp} used. Specially, according to the Facial Action Coding System (FACS) [139], Action Units (AUs) identify the fundamental muscle movements of the human face and can be considered as a proxy for the overall facial expressions. Bursic et al. [140] proposed to extract the AUs from the original and de-identified faces and then compute the root-mean-square error (RMSE) averaged on all AUs. For video de-identification tasks, the Pearson's Correlation Coefficient (PCC) is additionally computed to measure how well the AUs correlate on the temporal dimension.

- *Temporal Consistency*: For video de-identification, preserving temporal consistency is an essential metric for data utility. As mentioned above, the adapted FID metric could measure temporal consistency in synthesized videos. Meanwhile, empirically, the lack of temporal consistency can be observed as flickering and warping faces with changing identities. Thus in order to obtain a quantitative metric for temporal coherence, Balaji et al. [141] introduced the *Identity Invariance Score*, identity distance between every two subsequent frames averaged over the entire sample, with the postulation of correlation between temporal coherence and invariance of altered identity. We note that this method could be generalized to evaluate temporal consistency in other attributes.

In summary, the evaluation measures discussed in this subsection provide a thorough framework for assessing the performance of face de-identification algorithms. By outlining metrics for both privacy protection and utility, this compilation addresses the dual goal of obscuring identities while keeping essential visual information. These metrics not only quantify how well identities are concealed but also measure the quality and naturalness of the de-identified faces, covering aspects like detectability, landmark consistency, perceptual similarity, and retaining attributes. Also, for video de-identification, the focus on maintaining coherence and consistency over time further improves the comprehensive evaluation of these algorithms. Together, these diverse metrics help advance the development of robust and effective face de-identification techniques that strike an optimal balance between protecting privacy and preserving usefulness.

References

1. M.S. Ryoo, B. Rothrock, C. Fleming, H.J. Yang, Privacy-preserving human activity recognition from extreme low resolution (2016). arXiv preprint arXiv:1604.03196
2. L.D. Harmon, B. Julesz, Masking in visual recognition: effects of two-dimensional filtered noise. Science **180**(4091), 1194–1197 (1973)
3. Z. You, S. Li, Z. Qian, X. Zhang, Reversible privacy-preserving recognition, in *2021 IEEE International Conference on Multimedia and Expo (ICME)* (IEEE, Piscataway, 2021), pp. 1–6
4. J. Redmon, S. Divvala, R. Girshick, A. Farhadi, You only look once: Unified, real-time object detection, in *Proceedings of the IEEE Conference on Computer Vision and Pattern Recognition* (2016), pp. 779–788

5. S.J. Oh, R. Benenson, M. Fritz, B. Schiele, Faceless person recognition: Privacy implications in social media, in *European Conference on Computer Vision* (Springer, Berlin, 2016), pp. 19–35
6. R. McPherson, R. Shokri, V. Shmatikov, Defeating image obfuscation with deep learning (2016). arXiv preprint arXiv:1609.00408
7. N. Vishwamitra, B. Knijnenburg, H. Hu, Y.P. Kelly Caine et al., Blur vs. block: Investigating the effectiveness of privacy-enhancing obfuscation for images, in *Proceedings of the IEEE Conference on Computer Vision and Pattern Recognition Workshops* (2017), pp. 39–47
8. A. Melle, J.-L. Dugelay, Scrambling faces for privacy protection using background self-similarities, in *2014 IEEE International Conference on Image Processing (ICIP)* (IEEE, Piscataway, 2014), pp. 6046–6050
9. G. Letournel, A. Bugeau, V.-T. Ta, J.-P. Domenger, Face de-identification with expressions preservation, in *2015 IEEE International Conference on Image Processing (ICIP)* (IEEE, Piscataway, 2015), pp. 4366–4370
10. M.A. Rafique, M.S. Azam, M. Jeon, S. Lee, Face-deidentification in images using restricted Boltzmann machines, in *2016 11th International Conference for Internet Technology and Secured Transactions (ICITST)* (IEEE, Piscataway, 2016), pp. 69–73
11. L. Yuan, T. Ebrahimi, Image privacy protection with secure jpeg transmorphing. IET Signal Process. **11**(9), 1031–1038 (2017)
12. P. Chriskos, J. Munro, V. Mygdalis, I. Pitas, Face detection hindering, in *2017 IEEE Global Conference on Signal and Information Processing (GlobalSIP)* (IEEE, Piscataway, 2017), pp. 403–407
13. S. Dadkhah, M. Koeppen, S. Sadeghi, K. Yoshida, Bad AI: Investigating the effect of half-toning techniques on unwanted face detection systems, in *2018 9th IFIP International Conference on New Technologies, Mobility and Security (NTMS)* (IEEE, Piscataway, 2018), pp. 1–5
14. L. Fan, Image pixelization with differential privacy, in *Data and Applications Security and Privacy XXXII: 32nd Annual IFIP WG 11.3 Conference, DBSec 2018, Bergamo, Italy, July 16–18, 2018, Proceedings 32* (Springer, Berlin, 2018), pp. 148–162
15. L. Fan, Differential privacy for image publication, in *Theory and Practice of Differential Privacy (TPDP) Workshop*, vol. 1, no. 2 (2019), p. 6
16. L. Fan, Practical image obfuscation with provable privacy, in *2019 IEEE International Conference on Multimedia and Expo (ICME)* (IEEE, Piscataway, 2019), pp. 784–789
17. M.U. Saleem, D. Reilly, L. Fan, Dp-shield: Face obfuscation with differential privacy, in *Advances in Database Technology* (2022)
18. C. Liu, J. Yang, W. Zhao, Y. Zhang, J. Li, C. Mu, Face image publication based on differential privacy. Wirel. Commun. Mob. Comput. **2021**, 1–20 (2021)
19. R. Jiang, A. Bouridane, D. Crookes, M.E. Celebi, H.-L. Wei, Privacy-protected facial biometric verification using fuzzy forest learning. IEEE Trans. Fuzzy Syst. **24**(4), 779–790 (2015)
20. R. Jiang, S. Al-Maadeed, A. Bouridane, D. Crookes, M.E. Celebi, Face recognition in the scrambled domain via salience-aware ensembles of many kernels. IEEE Trans. Inf. Forens. Secur. **11**(8), 1807–1817 (2016)
21. E.M. Newton, L. Sweeney, B. Malin, Preserving privacy by de-identifying face images. IEEE Trans. Knowl. Data Eng. **17**(2), 232–243 (2005)
22. B. Driessen, M. Dürmuth, Achieving anonymity against major face recognition algorithms, in *Communications and Multimedia Security: 14th IFIP TC 6/TC 11 International Conference, CMS 2013, Magdeburg, Germany, September 25–26, 2013. Proceedings 14*. Springer , 18–33 (2013)
23. R. Gross, E. Airoldi, B. Malin, and L. Sweeney, "Integrating utility into face de-identification," in *International Workshop on Privacy Enhancing Technologies* (Springer, Berlin, 2005), pp. 227–242

24. R. Gross, L. Sweeney, F. De la Torre, S. Baker, Model-based face de-identification, in *2006 Conference on Computer Vision and Pattern Recognition Workshop (CVPRW'06)* (IEEE, Piscataway, 2006), pp. 161–161
25. R. Gross, L. Sweeney, F. De La Torre, S. Baker, Semi-supervised learning of multi-factor models for face de-identification, in *2008 IEEE Conference on Computer Vision and Pattern Recognition* (IEEE, Piscataway, 2008), pp. 1–8
26. J. Prinosil, P. Kriz, K. Riha, M.K. Dutta, A. Issac, Facial image de-identification using active appearance model, in *2017 International Conference on Emerging Trends in Computing and Communication Technologies (ICETCCT)* (IEEE, Piscataway, 2017), pp. 1–5
27. L. Meng, Z. Sun, A. Ariyaeeinia, K.L. Bennett, Retaining expressions on de-identified faces, in *2014 37th International Convention on Information and Communication Technology, Electronics and Microelectronics (MIPRO)* (IEEE, Piscataway, 2014), pp. 1252–1257
28. L. Meng, Z. Sun, Face de-identification with perfect privacy protection, in *2014 37th International Convention on Information and Communication Technology, Electronics and Microelectronics (MIPRO)* (IEEE, Piscataway, 2014), pp. 1234–1239
29. Z. Sun, L. Meng, A. Ariyaeeinia, Distinguishable de-identified faces, in *2015 11th IEEE International Conference and Workshops on Automatic Face and Gesture Recognition (FG)*, vol. 4 (IEEE, Piscataway, 2015), pp. 1–6
30. L. Meng, Z. Sun, O. Tejada Collado, Efficient approach to de-identifying faces in videos. IET Signal Process. **11**(9), 1039–1045 (2017)
31. H. Chi, Y.H. Hu, Facial image de-identification using identity subspace decomposition, in *2014 IEEE International Conference on Acoustics, Speech and Signal Processing (ICASSP)* (IEEE, Piscataway, 2014), pp. 524–528
32. T. Sim, L. Zhang, Controllable face privacy, in *2015 11th IEEE International Conference and Workshops on Automatic Face and Gesture Recognition (FG)*, vol. 4 (IEEE, Piscataway, 2015), pp. 1–8
33. X. Wang, C. Xiong, Q. Pei, Y. Qu, Expression preserved face privacy protection based on multi-mode discriminant analysis. Comput. Mater. Continua **57**, 107–121 (2018)
34. L. Du, M. Yi, E. Blasch, H. Ling, Garp-face: Balancing privacy protection and utility preservation in face de-identification, in *IEEE International Joint Conference on Biometrics* (IEEE, Piscataway, 2014), pp. 1–8
35. A. Jourabloo, X. Yin, X. Liu, Attribute preserved face de-identification, in *2015 International Conference on Biometrics (ICB)* (IEEE, Piscataway, 2015), pp. 278–285
36. H. Chi, Y.H. Hu, Face de-identification using facial identity preserving features, in *2015 IEEE Global Conference on Signal and Information Processing (GlobalSIP)* (IEEE, Piscataway, 2015), pp. 586–590
37. B. Meden, Z. Emersic, V. Struc, P. Peer, κ-same-net: Neural-network-based face deidentification, in *2017 International Conference and Workshop on Bioinspired Intelligence (IWOBI)* (IEEE, Piscataway, 2017), pp. 1–7
38. B. Meden, Ž. Emeršič, V. Štruc, P. Peer, k-same-net: k-anonymity with generative deep neural networks for face deidentification. Entropy **20**(1), 60 (2018)
39. S. Guo, S. Feng, Y. Li, S. An, H. Dong, Integrating diversity into neural-network-based face deidentification, in *2018 37th Chinese Control Conference (CCC)* (IEEE, Piscataway, 2018), pp. 9356–9361
40. Y.-L. Pan, M.-J. Haung, K.-T. Ding, J.-L. Wu, J.-S. Jang, K-same-siamese-gan: K-same algorithm with generative adversarial network for facial image de-identification with hyper-parameter tuning and mixed precision training, in *2019 16th IEEE International Conference on Advanced Video and Signal Based Surveillance (AVSS)* (IEEE, Piscataway, 2019), pp. 1–8
41. M.-H. Le, M.S.N. Khan, G. Tsaloli, N. Carlsson, S. Buchegger, AnonFACES: Anonymizing faces adjusted to constraints on efficacy and security, in *Proceedings of the 19th Workshop on Privacy in the Electronic Society* (2020), pp. 87–100
42. L. Chuanlu, W. Yicheng, C. Hehua, W. Shuliang, Utility preserved facial image de-identification using appearance subspace decomposition. Chinese J. Electron. **30**(3), 413–418 (2021)

43. Y. Jeong, J. Choi, S. Kim, Y. Ro, T.-H. Oh, D. Kim, H. Ha, S. Yoon, FICGAN: Facial identity controllable GAN for de-identification (2021). arXiv preprint arXiv:2110.00740
44. P. Samarati, L. Sweeney, Protecting privacy when disclosing information: k-anonymity and its enforcement through generalization and suppression (1998). Technical Report. SRI International Computer Science Laboratory.
45. L. Sweeney, k-anonymity: A model for protecting privacy. Int. J. Uncert. Fuzz. Knowl.-Based Syst. **10**(05), 557–570 (2002)
46. S.R. Ganta, S.P. Kasiviswanathan, A. Smith, Composition attacks and auxiliary information in data privacy, in *Proceedings of the 14th ACM SIGKDD International Conference on Knowledge Discovery and Data Mining* (2008), pp. 265–273
47. A. Basu, T. Nakamura, S. Hidano, S. Kiyomoto, k-anonymity: Risks and the reality, in *2015 IEEE Trustcom/BigDataSE/ISPA*, vol. 1 (IEEE, Piscataway, 2015), pp. 983–989
48. M. Wang, W. Deng, Deep face recognition: a survey. Neurocomputing **429**, 215–244 (2021)
49. I.J. Goodfellow, J. Shlens, C. Szegedy, Explaining and harnessing adversarial examples (2014). arXiv preprint arXiv:1412.6572.
50. S.-M. Moosavi-Dezfooli, A. Fawzi, O. Fawzi, P. Frossard, Universal adversarial perturbations, in *Proceedings of the IEEE Conference on Computer Vision and Pattern Recognition* (2017), pp. 1765–1773
51. Y. Dong, F. Liao, T. Pang, H. Su, J. Zhu, X. Hu, J. Li, Boosting adversarial attacks with momentum, in *Proceedings of the IEEE Conference on Computer Vision and Pattern Recognition* (2018), pp. 9185–9193
52. D.J. Miller, Z. Xiang, G. Kesidis, Adversarial learning targeting deep neural network classification: a comprehensive review of defenses against attacks. Proc. IEEE **108**(3), 402–433 (2020)
53. M. Sharif, S. Bhagavatula, L. Bauer, M.K. Reiter, Accessorize to a crime: Real and stealthy attacks on state-of-the-art face recognition, in *Proceedings of the 2016 ACM SIGSAC Conference on Computer and Communications Security* (2016), pp. 1528–1540
54. M. Sharif, S. Bhagavatula, L. Bauer, M.K. Reiter, A general framework for adversarial examples with objectives. ACM Trans. Privacy Secur. **22**(3), 1–30 (2019)
55. S. Komkov, A. Petiushko, AdvHat: Real-world adversarial attack on ArcFace face ID system, in *2020 25th International Conference on Pattern Recognition (ICPR)* (IEEE, Piscataway, 2021), pp. 819–826
56. J. Deng, J. Guo, N. Xue, S. Zafeiriou, Arcface: Additive angular margin loss for deep face recognition, in *Proceedings of the IEEE Conference on Computer Vision and Pattern Recognition* (2019), pp. 4690–4699
57. E. Chatzikyriakidis, C. Papaioannidis, I. Pitas, Adversarial face de-identification, in *2019 IEEE International Conference on Image Processing (ICIP)* (IEEE, Piscataway, 2019), pp. 684–688
58. A. Rozsa, E.M. Rudd, T.E. Boult, Adversarial diversity and hard positive generation, in *Proceedings of the IEEE Conference on Computer Vision and Pattern Recognition Workshops* (2016), pp. 25–32
59. A. Kurakin, I.J. Goodfellow, S. Bengio, Adversarial examples in the physical world, in *Artificial Intelligence Safety and Security* (Chapman and Hall/CRC, Boca Raton, 2018), pp. 99–112
60. S.J. Oh, M. Fritz, B. Schiele, Adversarial image perturbation for privacy protection a game theory perspective, in *2017 IEEE International Conference on Computer Vision (ICCV)* (IEEE, Piscataway, 2017), pp. 1491–1500
61. B. Liu, J. Xiong, Y. Wu, M. Ding, C.M. Wu, Protecting multimedia privacy from both humans and AI, in *2019 IEEE International Symposium on Broadband Multimedia Systems and Broadcasting (BMSB)* (IEEE, Piscataway, 2019), pp. 1–6
62. D. Deb, J. Zhang, A.K. Jain, Advfaces: Adversarial face synthesis, in *2020 IEEE International Joint Conference on Biometrics (IJCB)* (IEEE, Piscataway, 2020), pp. 1–10
63. Y. Dong, H. Su, B. Wu, Z. Li, W. Liu, T. Zhang, J. Zhu, Efficient decision-based black-box adversarial attacks on face recognition, in *Proceedings of the IEEE/CVF Conference on Computer Vision and Pattern Recognition* (2019), pp. 7714–7722

64. Y. Zhong, W. Deng, Towards transferable adversarial attack against deep face recognition. IEEE Trans. Inf. Forens. Secur. **16**, 1452–1466 (2020)
65. J. Zhang, J. Sang, X. Zhao, X. Huang, Y. Sun, Y. Hu, Adversarial privacy-preserving filter, in *Proceedings of the 28th ACM International Conference on Multimedia* (2020), pp. 1423–1431
66. X. Yang, Y. Dong, T. Pang, H. Su, J. Zhu, Y. Chen, H. Xue, Towards face encryption by generating adversarial identity masks, in *Proceedings of the IEEE/CVF International Conference on Computer Vision* (2021), pp. 3897–3907
67. S. Yang, T. Guo, Y. Wang, C. Xu, Adversarial robustness through disentangled representations. Proc. AAAI Conf. Artif. Intell. **35**(4), 3145–3153 (2021)
68. L. Yang, Q. Song, Y. Wu, Attacks on state-of-the-art face recognition using attentional adversarial attack generative network. Multimedia Tool. Appl. **80**, 855–875 (2021)
69. J. Yang, W. Zhang, J. Liu, J. Wu, J. Yang, Generating de-identification facial images based on the attention models and adversarial examples. Alexandr. Eng. J. **61**(11), 8417–8429 (2022)
70. Y. Zhong, W. Deng, Opom: Customized invisible cloak towards face privacy protection. IEEE Trans. Pattern Analy. Mach. Intell. **45**, 3590–3603 (2022)
71. Z.-A. Zhu, Y.-Z. Lu, C.-K. Chiang, Generating adversarial examples by makeup attacks on face recognition, in *2019 IEEE International Conference on Image Processing (ICIP)* (IEEE, Piscataway, 2019), pp. 2516–2520
72. S. Hu, X. Liu, Y. Zhang, M. Li, L. Y. Zhang, H. Jin, L. Wu, Protecting facial privacy: Generating adversarial identity masks via style-robust makeup transfer, in *Proceedings of the IEEE/CVF Conference on Computer Vision and Pattern Recognition* (2022), pp. 15014–15023
73. A. Shafahi, W.R. Huang, M. Najibi, O. Suciu, C. Studer, T. Dumitras, T. Goldstein, Poison frogs! targeted clean-label poisoning attacks on neural networks, in *Advances in Neural Information Processing Systems* (2018), pp. 6103–6113
74. C. Zhu, W.R. Huang, H. Li, G. Taylor, C. Studer, T. Goldstein, Transferable clean-label poisoning attacks on deep neural nets, in *International Conference on Machine Learning*, PMLR (2019), pp. 7614–7623
75. S. Shan, E. Wenger, J. Zhang, H. Li, H. Zheng, B.Y. Zhao, Fawkes: Protecting privacy against unauthorized deep learning models, in *29th {USENIX} Security Symposium ({USENIX} Security 20)* (2020), pp. 1589–1604
76. Y. Choi, M. Choi, M. Kim, J.-W. Ha, S. Kim, J. Choo, Stargan: Unified generative adversarial networks for multi-domain image-to-image translation, in *Proceedings of the IEEE Conference on Computer Vision and Pattern Recognition* (2018), pp. 8789–8797
77. G. Yang, N. Fei, M. Ding, G. Liu, Z. Lu, T. Xiang, L2m-GAN: Learning to manipulate latent space semantics for facial attribute editing, in *Proceedings of the IEEE/CVF Conference on Computer Vision and Pattern Recognition* (2021), pp. 2951–2960
78. Y. Yang, M.Z. Hossain, T. Gedeon, S. Rahman, S2fgan: Semantically aware interactive sketch-to-face translation, in *Proceedings of the IEEE/CVF Winter Conference on Applications of Computer Vision* (2022), pp. 1269–1278
79. S. Mosaddegh, L. Simon, F. Jurie, Photorealistic face de-identification by aggregating donors' face components, in *Computer Vision–ACCV 2014: 12th Asian Conference on Computer Vision, Singapore, November 1–5, 2014, Revised Selected Papers, Part III 12* (Springer, Berlin, 2015), pp. 159–174
80. T. Li, L. Lin, Anonymousnet: Natural face de-identification with measurable privacy, in *Proceedings of the IEEE Conference on Computer Vision and Pattern Recognition Workshops* (2019), pp. 0–0
81. N. Li, T. Li, S. Venkatasubramanian, t-closeness: Privacy beyond k-anonymity and l-diversity, in *2006 IEEE 23rd International Conference on Data Engineering* (IEEE, Piscataway, 2007), pp. 106–115
82. Y.-L. Pan, J.-C. Chen, J.-L. Wu, A multi-factor combinations enhanced reversible privacy protection system for facial images, in *2021 IEEE International Conference on Multimedia and Expo (ICME)* (IEEE, Piscataway, 2021), pp. 1–6

83. L. Zhai, Q. Guo, X. Xie, L. Ma, Y.E. Wang, Y. Liu, A3gan: Attribute-aware anonymization networks for face de-identification, in *Proceedings of the 30th ACM International Conference on Multimedia* (2022), pp. 5303–5313
84. J. Jam, C. Kendrick, K. Walker, V. Drouard, J.G.-S. Hsu, M.H. Yap, A comprehensive review of past and present image inpainting methods. Comput. Vision Image Understand. **203**, 103147 (2021)
85. Q. Sun, L. Ma, S. Joon Oh, L. Van Gool, B. Schiele, M. Fritz, Natural and effective obfuscation by head inpainting, in *Proceedings of the IEEE Conference on Computer Vision and Pattern Recognition* (2018), pp. 5050–5059
86. H. Hukkelås, R. Mester, F. Lindseth, Deepprivacy: A generative adversarial network for face anonymization, in *International Symposium on Visual Computing* (Springer, Berlin, 2019), pp. 565–578
87. Y. Qiu, Z. Niu, Q. Tian, B. Song, Privacy preserving facial image processing method using variational autoencoder, in *Big Data and Security: Third International Conference, ICBDS 2021, Shenzhen, November 26–28, 2021, Proceedings* (Springer, Berlin, 2022), pp. 3–17
88. Z. Kuang, H. Liu, J. Yu, A. Tian, L. Wang, J. Fan, N. Babaguchi, Effective de-identification generative adversarial network for face anonymization, in *Proceedings of the 29th ACM International Conference on Multimedia* (2021), pp. 3182–3191
89. H. Du, H. Shi, D. Zeng, X.-P. Zhang, T. Mei, The elements of end-to-end deep face recognition: a survey of recent advances. ACM Comput. Surv. **54**(10s), 1–42 (2022)
90. B. Meden, R.C. Mallı, S. Fabijan, H.K. Ekenel, V. Štruc, P. Peer, Face deidentification with generative deep neural networks. IET Signal Process. **11**(9), 1046–1054 (2017)
91. O.M. Parkhi, A. Vedaldi, A. Zisserman, Deep face recognition **1**(3), 6 (2015)
92. J. Chen, J. Konrad, P. Ishwar, Vgan-based image representation learning for privacy-preserving facial expression recognition, in *Proceedings of the IEEE Conference on Computer Vision and Pattern Recognition Workshops* (2018), pp. 1570–1579
93. P. Nousi, S. Papadopoulos, A. Tefas, I. Pitas, Deep autoencoders for attribute preserving face de-identification. Signal Process. Image Commun. **81**, 115699 (2020)
94. Z. Guo, H. Liu, Z. Kuang, Y. Nakashima, N. Babaguchi, Privacy sensitive large-margin model for face de-identification, in *Neural Computing for Advanced Applications: First International Conference, NCAA 2020, Shenzhen, July 3–5, 2020, Proceedings 1* (Springer, Berlin, 2020), pp. 488–501 (2020)
95. M. Gong, J. Liu, H. Li, Y. Xie, Z. Tang, Disentangled representation learning for multiple attributes preserving face deidentification. IEEE Trans. Neural Netw. Learn. Syst. **33**(1), 244–256 (2020)
96. Z. Kuang, Z. Guo, J. Fang, J. Yu, N. Babaguchi, J. Fan, Unnoticeable synthetic face replacement for image privacy protection. Neurocomputing **457**, 322–333 (2021)
97. O. Ronneberger, P. Fischer, T. Brox, U-net: Convolutional networks for biomedical image segmentation, in *Medical Image Computing and Computer-Assisted Intervention-MICCAI 2015, 18th International Conference, Munich, October 5–9, 2015, Proceedings, Part III 18* (Springer, Berlin, 2015), pp. 234–241
98. M. Maximov, I. Elezi, L. Leal-Taixé, Ciagan: Conditional identity anonymization generative adversarial networks, in *Proceedings of the IEEE/CVF Conference on Computer Vision and Pattern Recognition* (2020), pp. 5447–5456
99. M. Maximov, I. Elezi, L. Leal-Taixé, Decoupling identity and visual quality for image and video anonymization, in *Proceedings of the Asian Conference on Computer Vision* (2022), pp. 3637–3653
100. Y. Yang, Y. Huang, M. Shi, K. Chen, W. Zhang, Invertible mask network for face privacy preservation. Inf. Sci. **629**, 566–579 (2023)
101. R. Chen, X. Chen, B. Ni, Y. Ge, Simswap: An efficient framework for high fidelity face swapping, in *Proceedings of the 28th ACM International Conference on Multimedia* (2020), pp. 2003–2011
102. Y. Wu, F. Yang, Y. Xu, H. Ling, Privacy-protective-gan for privacy preserving face de-identification. J. Comput. Sci. Technol. **34**(1), 47–60 (2019)

103. J. Lin, Y. Li, G. Yang, Fpgan: face de-identification method with generative adversarial networks for social robots. Neural Netw. **133**, 132–147 (2021)
104. M.H. Khojaste, N.M. Farid, A. Nickabadi, GMFIM: a generative mask-guided facial image manipulation model for privacy preservation. Comput. Graph. **112**, 81–91 (2023)
105. T. Karras, S. Laine, T. Aila, A style-based generator architecture for generative adversarial networks, in *Proceedings of the IEEE/CVF Conference on Computer Vision and Pattern Recognition* (2019), pp. 4401–4410
106. Y. Zhao, B. Liu, T. Zhu, M. Ding, W. Zhou, Private-encoder: enforcing privacy in latent space for human face images. Concurr. Comput. Pract. Exp. **34**(3), e6548 (2022)
107. T. Karras, S. Laine, M. Aittala, J. Hellsten, J. Lehtinen, T. Aila, Analyzing and improving the image quality of stylegan, in *Proceedings of the IEEE/CVF Conference on Computer Vision and Pattern Recognition* (2020), pp. 8110–8119
108. D. Cho, J.H. Lee, I.H. Suh, Cleanir: Controllable attribute-preserving natural identity remover. Appl. Sci. **10**(3), 1120 (2020)
109. K. He, X. Zhang, S. Ren, J. Sun, Deep residual learning for image recognition, in *Proceedings of the IEEE Conference on Computer Vision and Pattern Recognition* (2016), pp. 770–778
110. Y. Luo, J. Zhu, K. He, W. Chu, Y. Tai, C. Wang, J. Yan, Styleface: Towards identity-disentangled face generation on megapixels, in *Computer Vision-ECCV 2022, 17th European Conference, Tel Aviv, October 23–27, 2022, Proceedings, Part XVI* (Springer, Berlin, 2022), pp. 297–312
111. Y. Huang, Y. Wang, Y. Tai, X. Liu, P. Shen, S. Li, J. Li, F. Huang, Curricularface: adaptive curriculum learning loss for deep face recognition, in *Proceedings of the IEEE/CVF Conference on Computer Vision and Pattern Recognition* (2020), pp. 5901–5910
112. K.H. Cheng, Z. Yu, H. Chen, G. Zhao, Benchmarking 3d face de-identification with preserving facial attributes, in *2022 IEEE International Conference on Image Processing (ICIP)* (IEEE, Piscataway, 2022), pp. 656–660
113. C. Szegedy, W. Liu, Y. Jia, P. Sermanet, S. Reed, D. Anguelov, D. Erhan, V. Vanhoucke, A. Rabinovich, Going deeper with convolutions, in *Proceedings of the IEEE Conference on Computer Vision and Pattern Recognition* (2015), pp. 1–9
114. K. Simonyan, A. Zisserman, Very deep convolutional networks for large-scale image recognition (2014). arXiv preprint arXiv:1409.1556
115. G. Huang, Z. Liu, L. Van Der Maaten, K.Q. Weinberger, Densely connected convolutional networks, in *Proceedings of the IEEE Conference on Computer Vision and Pattern Recognition* (2017), pp. 4700–4708
116. X. Gu, W. Luo, M.S. Ryoo, Y.J. Lee, Password-conditioned anonymization and deanonymization with face identity transformers,in *Computer Vision-ECCV 2020, 16th European Conference, Glasgow, August 23–28, 2020, Proceedings, Part XXIII 16* (Springer, Berlin, 2020), pp. 727–743
117. W. Liu, Y. Wen, Z. Yu, M. Li, B. Raj, L. Song, Sphereface: Deep hypersphere embedding for face recognition, in *Proceedings of the IEEE Conference on Computer Vision and Pattern Recognition* (2017), pp. 212–220
118. Q. Sun, A. Tewari, W. Xu, M. Fritz, C. Theobalt, B. Schiele, A hybrid model for identity obfuscation by face replacement, in *Proceedings of the European Conference on Computer Vision (ECCV)* (2018), pp. 553–569
119. W.L. Croft, J.-R. Sack, W. Shi, Differentially private obfuscation of facial images, in *Machine Learning and Knowledge Extraction: Third IFIP TC 5, TC 12, WG 8.4, WG 8.9, WG 12.9 International Cross-Domain Conference, CD-MAKE 2019, Canterbury, August 26–29, 2019, Proceedings 3* (Springer, Berlin, 2019), pp. 229–249
120. J.-W. Chen, L.-J. Chen, C.-M. Yu, C.-S. Lu, Perceptual indistinguishability-net (pi-net): Facial image obfuscation with manipulable semantics, in *Proceedings of the IEEE/CVF Conference on Computer Vision and Pattern Recognition* (2021), pp. 6478–6487
121. B. Liu, M. Ding, H. Xue, T. Zhu, D. Ye, L. Song, W. Zhou, Dp-image: Differential privacy for image data in feature space (2021). arXiv preprint arXiv:2103.07073

122. T. Li, C. Clifton, Differentially private imaging via latent space manipulation (2021). arXiv preprint arXiv:2103.05472
123. W.L. Croft, J.-R. Sack, W. Shi, Differentially private facial obfuscation via generative adversarial networks. Future Gener. Comput. Syst. **129**, 358–379 (2022)
124. M. Chu, Y. Xie, J. Mayer, L. Leal-Taixé, N. Thuerey, Learning temporal coherence via self-supervision for gan-based video generation. ACM Trans. Graph. **39**(4), 75–1 (2020)
125. B. Zhu, H. Fang, Y. Sui, L. Li, Deepfakes for medical video de-identification: Privacy protection and diagnostic information preservation, in *Proceedings of the AAAI/ACM Conference on AI, Ethics, and Society* (2020), pp. 414–420
126. B. Samarzija, S. Ribaric, An approach to the de-identification of faces in different poses, in *2014 37th International Convention on Information and Communication Technology, Electronics and Microelectronics (MIPRO)* (IEEE, Piscataway, 2014), pp. 1246–1251
127. Y. Li, S. Lyu, De-identification without losing faces, in *Proceedings of the ACM Workshop on Information Hiding and Multimedia Security* (2019), pp. 83–88
128. O. Gafni, L. Wolf, Y. Taigman, Live face de-identification in video, in *Proceedings of the IEEE/CVF International Conference on Computer Vision* (2019), pp. 9378–9387
129. R. Gross, L. Sweeney, J. Cohn, F. De la Torre, S. Baker, Face de-identification, in *Protecting Privacy in Video Surveillance* (Springer, Berlin, 2009), pp. 129–146
130. Z. Ren, Y.J. Lee, M.S. Ryoo, Learning to anonymize faces for privacy preserving action detection, in *Proceedings of the European Conference on Computer Vision (ECCV)* (2018), pp. 620–636
131. M. Xuan, J. Jiang, Video security algorithm aiming at the need of privacy protection, in *2009 Sixth International Conference on Fuzzy Systems and Knowledge Discovery*, vol. 5 (IEEE, Piscataway, 2009), pp. 473–477
132. P. Korshunov, T. Ebrahimi, Towards optimal distortion-based visual privacy filters, in *2014 IEEE International Conference on Image Processing (ICIP)* (IEEE, Piscataway, 2014), pp. 6051–6055
133. P. Korshunov, T. Ebrahimi, Using warping for privacy protection in video surveillance, in *2013 18th International Conference on Digital Signal Processing (DSP)* (IEEE, Piscataway, 2013), pp. 1–6
134. P. Korshunov, T. Ebrahimi, Using face morphing to protect privacy, in *2013 10th IEEE International Conference on Advanced Video and Signal Based Surveillance* (IEEE, Piscataway, 2013), pp. 208–213
135. K. Zhang, Z. Zhang, Z. Li, Y. Qiao, Joint face detection and alignment using multitask cascaded convolutional networks. IEEE Signal Process. Lett. **23**(10), 1499–1503 (2016)
136. R. Zhang, P. Isola, A.A. Efros, E. Shechtman, O. Wang, The unreasonable effectiveness of deep features as a perceptual metric, in *Proceedings of the IEEE Conference on Computer Vision and Pattern Recognition* (2018), pp. 586–595
137. M. Heusel, H. Ramsauer, T. Unterthiner, B. Nessler, S. Hochreiter, Gans trained by a two time-scale update rule converge to a local nash equilibrium, in *Advances in Neural Information Processing Systems*, vol. 30 (2017)
138. T.-C. Wang, M.-Y. Liu, J.-Y. Zhu, G. Liu, A. Tao, J. Kautz, B. Catanzaro, Video-to-video synthesis (2018). arXiv preprint arXiv:1808.06601
139. Y.-I. Tian, T. Kanade, J.F. Cohn, Recognizing action units for facial expression analysis. IEEE Trans. Pattern Analy. Mach. Intell. **23**(2), 97–115 (2001)
140. S. Bursic, A. D'Amelio, M. Granato, G. Grossi, R. Lanzarotti, A quantitative evaluation framework of video de-identification methods, in *2021 25th International Conference on Pattern Recognition (ICPR)* (IEEE, Piscataway, 2021), pp. 6089–6095
141. T. Balaji, P. Blies, G. Göri, R. Mitsch, M. Wasserer, T. Schön, Temporally coherent video anonymization through gan inpainting (2021). arXiv preprint arXiv:2106.02328
142. J. Cao, B. Liu, Y. Wen, R. Xie, L. Song, Personalized and invertible face de-identification by disentangled identity information manipulation, in *Proceedings of the IEEE/CVF International Conference on Computer Vision* (2021), pp. 3334–3342

Chapter 4
Face Image Privacy Protection with Differential Private k-Anonymity

4.1 Introduction

The major challenge of face de-identification is the tradeoff between privacy and image utility. The ideal de-identification method should be able to control the balance to adapt to extensive applications. Most GAN-based methods fail to quantify this matter until Li et al. [1] proposed that facial privacy is measurable and provided a privacy preservation way with an attribute selection method based on privacy metrics such as k-anonymity [2], l-diversity [3], and t-closeness [4]. However, AnonymousNet modified facial attributes of protected image close to its real-world distribution without considering the control of image disturbance degree. We hope to add minor changes for better utility preservation with the condition of privacy protection.

In this chapter, we propose a face image privacy protection method with differential private k-anonymity, including the following two key features: (1) it first finds the average face attributes of the k nearest neighbors of the given image and then edits it toward the direction of the average face, which can hide the identity while ensuring the modification is small. (2) Differential privacy (DP) is introduced to add randomness and provides further protection on top of the former because the first step is a deterministic process and limited in protection effectiveness. As the de-identification results shown in Fig. 4.1, our approach can generate the naturally realistic faces and keep similarity with the original images.

Main contents of this chapter have been published in "Cao, J., Liu, B., Wen, Y., Zhu, Y., Xie, R., Song, L., ... & Yin, Y. (2022, June). Hiding among your neighbors: Face image privacy protection with differential private k-anonymity. In 2022 IEEE International Symposium on Broadband Multimedia Systems and Broadcasting (BMSB) (pp. 1–6). IEEE."

Y. Wen et al., *Face De-identification: Safeguarding Identities in the Digital Era*,
https://doi.org/10.1007/978-3-031-58222-6_4

Fig. 4.1 Qualitative comparison of various traditional and state-of-the-art (SOTA) de-identification methods, where (**a**) input image, (**b**)–(**d**) are traditional methods including (**b**) blurred image, (**c**) adding Gaussian noise to the pixel, (**d**) pixelated image, and (**e**)–(**g**) are GAN-based methods including (**e**) DeepPrivacy [5], (**f**) CIAGAN [6], (**g**) AnonymousNet [1], and (**h**) de-identified results by our approach. As can be seen from the figures, the proposed method results in more natural facial images and can retain some similarity with the original

The major contributions of this chapter can be summarized in threefolds:

1. We propose an image de-identification method to achieve metric privacy based on attributes indistinguishability.
2. We design the differential private k-anonymity algorithm to select obfuscation attributes that can control the tradeoff between privacy and utility by adjustable parameters.
3. We empirically evaluate our obfuscation method for both privacy and utility performance. We show our approach can achieve higher utility and more similarity with the original face while ensuring privacy protection.

4.2 Related Works

4.2.1 Privacy-Preserving Machine Learning

The focus of privacy-preserving machine learning is how to prevent leaking sensitive information in both models and datasets. Plentiful researches are from the perspective of the target and stage of attacks, including membership inference [7], feature estimation [8], model-inversion attacks [9], etc. Differential private machine learning has been widely used in perturbation, which aims to train models with formal guarantees implemented by randomizing the training process such as adding noise to the gradient. The advanced interaction between differential privacy and machine learning has been discussed in [10]. Along with the advancement of collaborative learning, more researchers pay attention to privacy in the aggression process. A model-agnostic approach named "Private Aggregation of Teacher Ensembles" (PATE) [11] introduces a model aggregation strategy that injects randomness in the aggregation process to achieve privacy protection. Another more data-efficient algorithm named Private kNN [12] is the first practical differentially private deep learning solution for large-scale computer vision that can achieve comparable or better consequences than PATE while reducing privacy loss. Inspired by privacy-preserving strategies, we design the obfuscation algorithm in the attributes aggression process and apply it to the face de-identification task.

4.2.2 GAN-Based Face Manipulation

Generative Adversarial Networks (GANs) [13] is originally proposed to generate images from random noise, which have shown remarkable results in various computer vision tasks including image generation [14], image translation [15], face image synthesis [16], and so on. Facial attribute editing can be regarded as a multidomain image-to-image translation problem that has received extensive attention and research. Conditional GAN (cGAN) [17] considers class information in the training process of generator and discriminator to generate samples conditioned on the desired class. IcGAN [18] adopts an encoder to generate latent code and a cGAN to decode it conditioned on target attributes. AttGAN [19] applies an attribute classification constraint and takes the target attribute vector as input to the transform model. In StarGAN [20], the multidomain translation problem is approached using a single generator instead of a separate generator for each domain. STGAN [21] is developed from AttGAN by presenting selective transfer units incorporated with encoder–decoder. Li et al. [22] proposed the hierarchical style disentanglement (HiSD) to avoid uncontrolled global manipulations in image translation. In our approach, after obtaining the obfuscation attributes, we reconstruct the de-identification results by a face editing GAN to generate the natural facial image with pleasant visual perception.

4.3 Preliminaries

In this section, we present the problem formulation for face de-identification task and a brief introduction to the necessary technical components in our approach.

4.3.1 Differential Privacy

Differential privacy is a rigorous mathematical definition of privacy and probability is used to take over randomness, which is a strong guarantee since it is based on the statistical property of the mechanism without the requirement of auxiliary information [23].

Definition 4.1 (ε-Differential Privacy) Let ε be a positive real number (privacy parameter), and the randomized algorithm $\mathcal{A} : Y \rightarrow \Theta$ is said to provide ε-differential privacy if for all neighboring datasets $D, D' \in Y$ that differ on at most a single element, and all random subsets $S \subset \Theta$ satisfy:

$$\Pr[\mathcal{A}(D) \in S] \leq e^{\varepsilon} \cdot \Pr\left[\mathcal{A}\left(D'\right) \in S\right]. \tag{4.1}$$

There are three commonly used mechanisms in differential privacy according to data types: *Laplace*, *Gaussian*, and *exponential mechanism*. The overall idea of the exponential mechanism is that when receiving a query, it returns a certain probability value calculated by the scoring function q instead of a deterministic result, thereby achieving differential privacy.

Definition 4.2 (Exponential Mechanism) Let $q(D, r)$ be a function of dataset D which selects and outputs an element $r \in R$, and then an exponential mechanism $\mathcal{M}$ is ε-differential privacy if

$$\mathcal{M}(D) = \left\{ \text{return } r \text{ with probability} \propto \exp\left(\frac{\varepsilon q(D, r)}{2\Delta q}\right)\right\}, \tag{4.2}$$

where Δq represents the sensitivity of function q.

4.3.2 Privacy Amplification

Subsampling is a widely used tool for privacy amplification, and Poisson sampling is the process where each data is subjected to an independent Bernoulli trial to be chosen with probability γ. When we apply the ε-differential privacy mechanism to a random γ-proportion subset, the whole procedure satisfies $O(\gamma\varepsilon)$-differential privacy, which is known as "subsampling lemma" or "secrecy of the samples" in the literature [24]. In our approach, Poisson sampling is applied for privacy amplification that can provide a stronger privacy guarantee.

4.4 Our Approach

We will describe our three-step approach in detail in this section. First of all, we employ the facial attribute classifier to predict original attributes. Then we calculate the obfuscation attributes with differential private k-anonymity algorithm. Finally, we employ the face attribute editing network to generate de-identification results.

4.4.1 *Step 1: Attributes Prediction*

Firstly, we train a facial attribute extraction network to predict labels of query X, which has two major functions in subsequent operations. On the one hand, when calculating the obfuscation attributes, it will be used as feature extractor to get deep features. On the other hand, when generating the de-identification, we take the different attributes as input, so we need the original prediction for reference.

For the c-label classification problem, we adopt MultiLabelSoftMarginLoss as loss function, which creates a criterion that optimizes a multilabel one-versus-all loss based on max entropy. For each sample in the minibatch:

$$\begin{aligned}\mathcal{L}(u, v) = & -\frac{1}{c}\sum_{i}^{c} v[i]\log\left((1+\exp(-u[i]))^{-1}\right) \\ & + (1-v[i])\log\left(\frac{\exp(-u[i])}{1+\exp(-u[i])}\right),\end{aligned} \tag{4.3}$$

where $v[i] \in \{0, 1\}$. Prediction label u and ground truth v are with the same shape of (n, c), where n is the batch size, while c represents the number of classes. At the end of this step, we can get the original attributes $\mathbb{P}$ of the given image.

4.4.2 *Step 2: Obfuscation*

We design differential private k-anonymity algorithm shown in Algorithm 1 to acquire the obfuscation attributes, which can be summarized as the following two parts, and we will further describe their respective functions in Sect. 4.5.3.

4.4.2.1 k-Anonymity Average Attributes

For the given no-label query X, we sample a random subset D_γ with the Poisson sampling of probability γ. Both X and D_γ will be mapped into the feature space by

Algorithm 1: Differential private k-anonymity algorithm

Input: Given image X, the ratio of Poisson sampling γ, the number of nearest neighbors participating in voting k, the parameter of differential privacy ε.
Output: Obfuscation Attributes $\mathbb{O}$.

```
// k-anonymity Prediction
Downsample the training set to obtain the random subset D_γ with Poisson sampling of
probability γ.
Map X and D_γ to the feature space and select k nearest neighbors from D_γ based on feature
distance.
Calculate the k-anonymity average attributes ℝ and voting results 𝕍 = {v_1, v_2, ..., v_n}.
// Independent Attributes Set
for attribute a_i in independent attributes 𝔹 do
    Calculate the probability p of with a_i by Eq. (4.5).
    if rand(0, 1) <p then
        Add with attribute a_i to 𝕆.
    else
        Add without attribute a_i to 𝕆.
    end if
end for
// Conflict Attributes Set
for group 𝔾_i in conflict attributes ℂ do
    Calculate each probability p_{i_n} corresponding to attributes a_{i_n} in 𝔾_i = {a_{i_1}, a_{i_2}, ..., a_{i_m}}
    according to Eq. (4.6).
    if rand(0, 1) ∈ [Σ_{j=1}^{k-1} p_{i_j}, Σ_{j=1}^{k} p_{i_j}] then
        Add with attribute a_{i_k}, without attributes a_{i_{n(n≠k)}} to 𝕆.
    end if
end for
return Obfuscation Attributes 𝕆.
```

a pretrained feature extractor φ. Then we select k nearest neighbors according to the feature Euclidean distance between $x = \varphi(X)$ and $f = \left\{f_i = \varphi(d_i) \mid \forall d_i \in D_\gamma\right\}$. Notice that for a binary classification task, the global sensitivity is 2, while for a problem with c-labels, the global sensitivity will be extended to $2c$, which will make the following noisy-adding mechanisms inefficient. In order to limit the range of global sensitivity, we apply τ-approximation [12] limitation that means each neighbor can only vote for τ attributes at most.

Definition 4.3 (τ-Approximation) Considering the binary multilabel task, the vote of neighbor j upon query X can be expressed as a c-way vector, and we apply

$$\hat{v}_{j,i} = v_{j,i} \cdot \min\left(\frac{\tau}{\mid v_j(X) \mid}, 1\right), i \in [1, c], \tag{4.4}$$

where $|v_j(x)|$ is the $\mathcal{L}_1$-norm of original neighbor j's voting results and $\hat{v}_j$ is the neighbor j's voting results with τ-approximation. The global sensitivity of a randomized algorithm $\mathcal{M}_\tau$ can be reduced to 2τ with this setting.

4.4.2.2 Differential Privacy

After obtaining the k-anonymity average attributes $\mathbb{R}$ and the corresponding votes $\mathbb{V} = \{v_1, v_2, \ldots, v_c\}$. To introduce more randomness for privacy protection, we further apply *exponential differential privacy* to the voting process as privacy metrics. We divide all considered privacy-sensitive attributes $\mathbb{A}$ into *independent attributes* $\mathbb{B}$ and *conflict attributes* $\mathbb{C}$, which satisfy $\mathbb{A} = \mathbb{B} \cup \mathbb{C}$ and $\mathbb{B} \cap \mathbb{C} = \varnothing$:

- Independent Attributes $\mathbb{B}$. There is no correlation between independent attribute a_i and other attributes in $\mathbb{B}$, that is, we can individually determine whether to choose it. Therefore, we count the voting results *with* this attribute v_i as the value of score function q and select the obfuscation attributes based on the probability calculated by

$$p = \frac{\exp\left(\frac{\varepsilon v_i}{2\Delta q}\right)}{\exp\left(\frac{\varepsilon v_i}{2\Delta q}\right) + \exp\left(\frac{\varepsilon(k-v_i)}{2\Delta q}\right)}. \tag{4.5}$$

- Conflict Attributes $\mathbb{C}$. Considering the exclusivity between attributes, we further divide conflict attributes set $\mathbb{C}$ into groups as $\mathbb{C} = \{\mathbb{G}_1, \mathbb{G}_2, \ldots, \mathbb{G}_m\}$, where there is no mutual influence between different groups. Generally speaking, two or more attributes in the same group $\mathbb{G}_i$ will not be selected simultaneously. We respectively count the votes of each attribute as the score function q value and the probability of selecting attribute a_{i_n} in $\mathbb{G}_i = \{a_{i_1}, a_{i_2}, \ldots, a_{i_m}\}$ by

$$p_{i_n} = \frac{\exp\left(\frac{\varepsilon v_{i_n}}{2\Delta q}\right)}{\sum_{j=1}^{m} \exp\left(\frac{\varepsilon v_{i_j}}{2\Delta q}\right)}. \tag{4.6}$$

4.4.3 Step 3: Image Generation

We adopt a GAN to generate de-identification images according to the obfuscation attributes. For better generation and feature accuracy, we customize the facial attribute editing model based on STGAN [21] that improves manipulation ability by presenting selective transfer units incorporated with encoder–decoder. Different from StarGAN [20] and AttGAN [19], which both take target attributes as input, STGAN only focuses on the changed attributes $attr_{\text{diff}}$ that represent the difference between predicted original facial attributes $\mathbb{P}$ and the obfuscation attributes $\mathbb{O}$ in our approach.

The loss function includes adversarial loss $\mathcal{L}_{adv}$, reconstruction loss $\mathcal{L}_{rec}$, and attribute manipulation loss $\mathcal{L}_{attr}$. The adversarial loss [13] is applied for constraining the generated results to be indistinguishable from real images. We follow Wasserstein GAN (WGAN) and WGAN-GP [25] to define the adversarial loss as

$$\max_{D_{adv}} \mathcal{L}_{D_{adv}} = \mathbb{E}_X D_{adv}(X) - \mathbb{E}_{\hat{Y}} D_{adv}(\hat{Y}) + \lambda \mathbb{E}_{\hat{X}} \left[\left(\left\| \nabla_{\hat{X}} D_{adv}(\hat{X}) \right\|_2 - 1 \right)^2 \right], \tag{4.7}$$

$$\max_{G} \mathcal{L}_{G_{adv}} = \mathbb{E}_{X, attr_{\text{diff}}} D_{adv} \left(G \left(X, attr_{\text{diff}} \right) \right), \tag{4.8}$$

where $\hat{X}$ is uniformly sampled between a pair of original and generated images and $\hat{Y} = G\left(X, attr_{\text{diff}}\right)$.

The reconstruction loss is defined as

$$\mathcal{L}_{rec} = \|X - G(X, 0)\|_1, \tag{4.9}$$

where the $\mathcal{L}_1$-norm distance is adopted for ensuring the quality and clarity of the reconstructed images and $G(X, 0)$ is the reconstructed images sharing the same attributes with the original.

To improve the accuracy of attributes editing, we introduce the attribute manipulation loss $\mathcal{L}_{attr}$. The attribute classifier D_{attr} shares the common convolution layers with D_{adv}, and the attribute manipulation loss is designed as

$$\mathcal{L}_{D_{attr}} = -\sum_{i=1}^{c} \left[attr_p^{(i)} \log D_{attr}^{(i)}(X) + \left(1 - attr_p^{(i)}\right) \log \left(1 - D_{attr}^{(i)}(X)\right) \right], \tag{4.10}$$

$$\mathcal{L}_{G_{attr}} = -\sum_{i=1}^{c} \left[attr_o^{(i)} \log D_{attr}^{(i)}(\hat{Y}) + \left(1 - attr_o^{(i)}\right) \log \left(1 - D_{attr}^{(i)}(\hat{Y})\right) \right], \tag{4.11}$$

where $attr_p^{(i)}$ means the i-th value of prediction attributes $\mathbb{P}$, $attr_o^{(i)}$ indicates the i-th value of obfuscation attributes $\mathbb{O}$, and $D_{attr}^{(i)}(X)$ represents the i-th value of attribute classification results of X by the attribute classifier D_{attr}.

Taking the above losses into account, the overall loss function of discriminator D can be formulated as

$$\mathcal{L}_D = -\mathcal{L}_{D_{adv}} + \lambda_1 \mathcal{L}_{D_{attr}}, \tag{4.12}$$

and that for the generator G is

$$\mathcal{L}_G = -\mathcal{L}_{G_{adv}} + \lambda_2 \mathcal{L}_{G_{attr}} + \lambda_3 \mathcal{L}_{rec}, \tag{4.13}$$

where λ_1, λ_2, and λ_3 are the model tradeoff parameters.

4.5 Experiments

4.5.1 *Dataset*

We use large-scale CelebFaces Attributes (CelebA) Dataset [26] that contains 202,599 aligned facial images and 10,177 identities with 40 *with or without* attributes labels of boolean values. In experiments, we use about half of the dataset, of which 75,160 images for training and 26,216 images for test. All images are firstly aligned to 178 × 218, and after detecting face region by dlib packages, we crop the images to the size of 128 × 128.

4.5.2 *Implementation Details*

Attributes Prediction We train the facial attributes classification network on CelebA dataset using the Resnet-50 structure. We conduct the batch size of 128, set a base learning rate of 4×10^{-4} reducing by a polynomial decay with a gamma of 0.1, and the weight decay is 5×10^{-4}.

Attributes Obfuscation When performing de-identification for the given image, we firstly downsample the training set in proportion to γ to get a random subset D_γ and then extract deep features from the fully connected layers of the facial attributes classification network. In our experiments, we consider 13 attributes to protect, including *Bald, Bangs, Black Hair, Blond Hair, Brown Hair, Bushy Eyebrows, Eyeglasses, Male, Mouth Slightly Open, Mustache, No Beard, Pale Skin*, and *Young*, due to that they are more distinctive in appearance. Among the attributes considered, we define two sets of conflicting attributes: $\mathbb{G}_1$ = {*Black Hair, Blond Hair, Brown Hair*} and $\mathbb{G}_2$ = {*Mustache, No Beard*}, while the others are all defined as independent attributes.

Image Generation Network We utilize the facial attributes editing to generate de-identified images after obtaining the obfuscation attributes. We train on CelebA dataset for the considered attributes following the settings in [21], where the tradeoff parameters in Eqs. (4.12) and (4.13) are set to $\lambda_1 = 1$, $\lambda_2 = 10$, and $\lambda_3 = 100$.

4.5.3 *Performance Analysis*

Here we present further analysis of our methods including perception effects, evaluation of *differential private k-anonymity* algorithm, and the influence of major parameters on experimental results.

Figure 4.2 illustrates some de-identification results in pairs, where the left presents the original image and the right is the de-identified result generated by our

Fig. 4.2 Some de-identification results generated by our approach. In each pair, the left is the original image, while the right is the de-identified

approach. To further explain the necessity of the two main parts (k-anonymity and differential privacy) in Algorithm 1, we compare the obfuscation extent of whether introducing differential privacy under different k values. We use the *attributes accuracy* between two attributes sets $\mathbb{M}$ and $\mathbb{N}$ as metrics, which is defined as

$$Accuracy = \frac{card(\mathbb{M}) - d(\mathbb{M}, \mathbb{N})}{card(\mathbb{M})}, \tag{4.14}$$

where $card(\mathbb{M})$ represents the number of elements in set $\mathbb{M}$, which is equal to that of $\mathbb{N}$, and $d(\mathbb{M}, \mathbb{N})$ represents the hamming distance between $\mathbb{M}$ and $\mathbb{N}$. Higher accuracy indicates a smaller difference.

We set $\Delta q = 1$, $\varepsilon = [0, 0.3]$ with a step length of 0.001 and Poisson subsampling ratio $\gamma = 0.05$ to respectively calculate the attributes accuracy with different k values between (a) k-anonymity average attributes and prediction (the green line), (b) *differential private k-anonymity* obfuscation attributes and prediction (the red line), (c) *differential private k-anonymity* obfuscation attributes and k-anonymity average attributes (the blue line). From the results shown in Fig. 4.3, we can conclude that there is a high attributes accuracy between k-anonymity

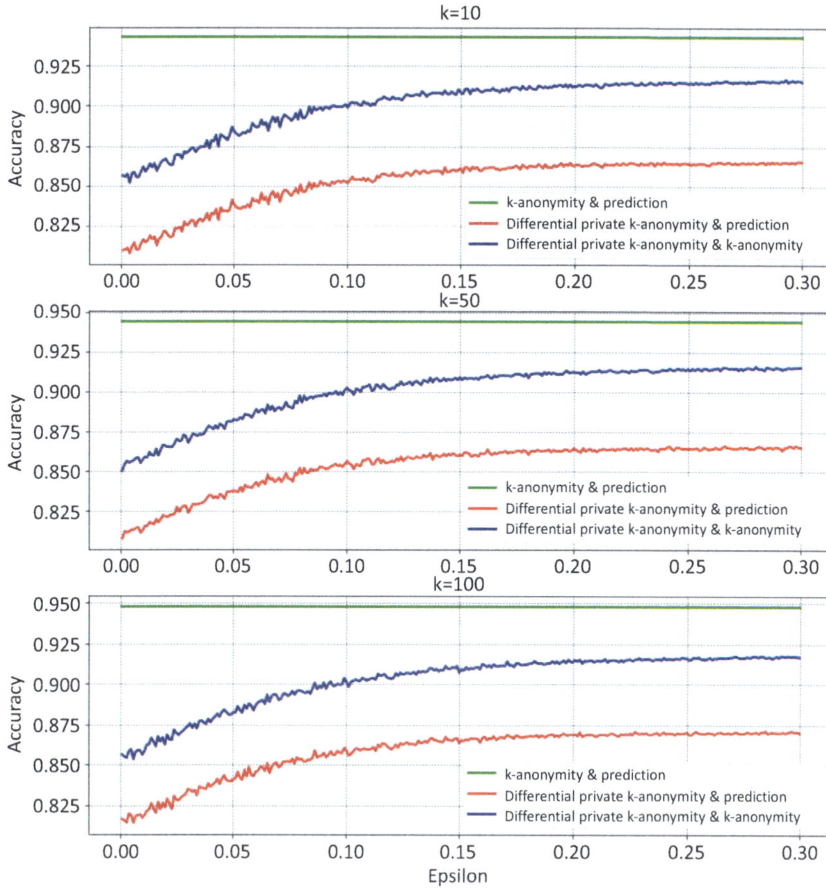

Fig. 4.3 The impact of differential privacy and k-anonymity on attributes accuracy, where the green line represents the effect of k-anonymity, the blue line indicates the randomness introduced by differential privacy, and the red line presents the total effect

average labels $\mathbb{R}$ and the predicted labels $\mathbb{P}$, which can provide the overall control to minimize modification in the de-identification process. However, the pure apply of k-anonymity fails to protect sensitive information from the homogeneity attack and is vulnerable to the attacks based on background knowledge [27]. Moreover, the protection effectiveness is limited especially when the value of k is large. Therefore, we employ differential privacy to provide more randomness in obfuscation process of more reliable privacy guarantees.

The influence of two main parameters k and ε on the attribute obfuscation is shown in Fig. 4.4, where the accuracy displayed on the y-axis is represented between obfuscation attributes $\mathbb{O}$ and prediction attributes $\mathbb{P}$. We only perturb the considered attributes, while the other attributes without privacy protection keep the same as the predicted. When we set $\varepsilon = 0.0$, it means randomly selecting either *with* or *without*

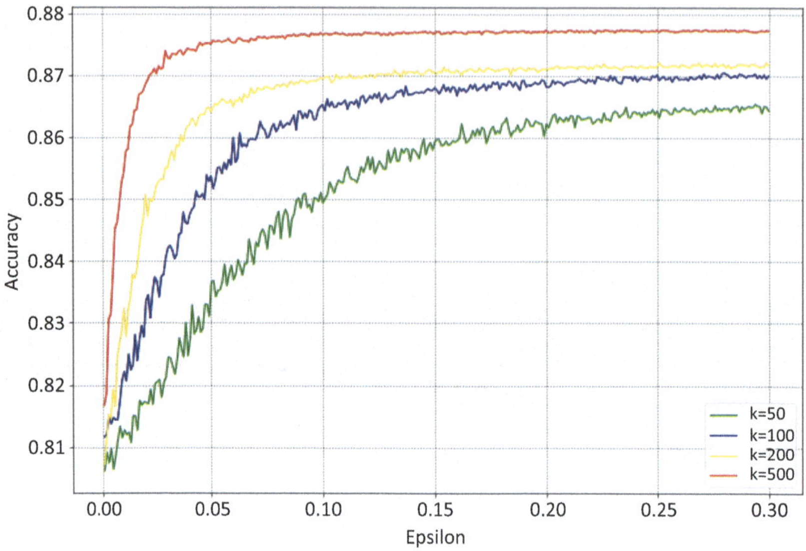

Fig. 4.4 The influence of different k and ε values on the obfuscation degree, where the y-axis represents *attributes accuracy* between obfuscation attributes and the prediction

independent attributes and choosing one of the conflict attributes in same group, both with the same probability. As ε increases, the extent of disturbance decreases, and the accuracy will increase. In addition, the attributes accuracy will be greater as k increases with the same ε, and the impact of k values has been magnified after the introduction of differential privacy because function q mainly depends on the voting results. Particularly, due to the design of conflicting attributes mechanism and the prediction deviation of k-anonymity, it will eventually stabilize instead of reaching 100%.

4.5.4 *Quantitative Evaluation*

We use the following metrics to evaluate our approach comparing with existing de-identification methods from both identity protection effectiveness and image utility:

- **Identity protection effectiveness**:
  - **Identity Distance (*Id-dis*)**. Most of face verification models judge whether two images have the same identity by comparing identity embedding distance. We use the Face Recognition to calculate the **identity distance (*Id-dis*)**, which is based on the deep learning model of dlib, and the model tested with Labelled Faces in the Wild datasets can achieve the accuracy of 99.38%.

- **Image utility**:
  - **Image quality**:
    - PSNR and SSIM. We use peak signal-to-noise ratio and structure similarity to measure image distortion at the pixel level. It should be noticed that these indicators just focus on objective image quality evaluation and fail to capture many nuances of human perception [28].
    - FID. We use Fréchet Inception Distance [29] to measure image distance in feature space calculated by the Inception-V3 model. When applying system distortion, lower FID indicates higher image quality.
    - LPIPS. Learned perceptual image patch similarity [30] distance is applied to measure visual similarity that is closer to human perception than traditional metrics.
  - **Utility for computer vision tasks**:
    - **Face Detectability (*Face-det*)**. We evaluate whether the de-identification results are still usable for identity-independent computer vision tasks by performing face detection with *opencv*. We define the proportion of faces can still be detected in the protected images as face detectability.

The comparison results are presented in Table 4.1. Since the strict threshold for judging whether two images have the same identity is 0.5 in the face recognition model, we choose the values of k and ε to make *Id-dis* basically meet the threshold. We select two sets of parameters with a smaller obfuscation and a larger in traditional methods including blurring, noise, and pixelation, and it can be concluded that when adding a small disturbance, there is little impact on image quality but almost no effects on identity protection. Increasing the degree of disturbance contributes higher protection effectiveness, but the image quality and utility will be damaged greatly. Compared with traditional methods, the GAN-

Table 4.1 Comparison with other methods under different metrics

	Id-dis↑	PSNR↑	SSIM↑	FID↓	LPIPS↓	Face-det↑
Blurring(r = 5)	0.2573	24.931	0.8005	66.866	0.0654	0.8600
Blurring(r = 20)	0.4203	22.666	0.7419	91.623	0.0755	0.6917
Noise(σ = 10)	0.2565	21.917	0.7739	32.126	0.0534	0.8136
Noise(σ = 30)	0.2911	17.968	0.6281	83.169	0.1265	0.2832
Pixelation(4 × 4)	0.3251	**25.221**	**0.8278**	**26.073**	0.0326	0.9302
Pixelation(8 × 8)	0.6908	22.686	0.7010	83.666	0.0915	0.0211
DeepPrivacy [5]	**0.7232**	20.046	0.7605	27.569	0.0868	0.9606
CIAGAN [6]	0.5740	19.014	0.5349	36.719	0.0782	0.9455
AnonymousNet [1]	0.4891	19.102	0.7380	55.047	0.0965	0.8224
Ours(k = 100,ε = 0.05)	0.5608	19.069	0.7726	52.888	0.0411	**0.9728**
Ours(k = 100,ε = 0.10)	0.5269	20.308	0.7588	52.214	0.0345	0.9614
Ours(k = 200,ε = 0.05)	0.4795	21.029	0.8024	38.315	**0.0323**	0.9502

based methods can balance the tradeoff better. Additionally, compared with the de-identification methods based on entire face synthesis like DeepPrivacy, our algorithm takes the reduction of modification degree into consideration, so that de-identified results can maintain higher perception similarity (lower LPIPS) with the original. CIAGAN is the identity-swapping-based anonymization method so that the de-identified face still corresponds to a piece of real identity information, which may cause identity leakage in the dataset. Due to the requirement of face landmarks and masked background in CIAGAN, it is not convenient in practical applications. Our approach can adjust the privacy protection level by controlling the parameters k and ε, so as to meet the application in different scenarios.

4.6 Conclusion

In this chapter, we focus on the problem of image privacy and face de-identification. In order to confuse the identity information with minor modifications, we propose a face image privacy protection method to provide metric privacy based on attributes indistinguishability. Our approach consists of three steps: attributes prediction, privacy protection attributes obfuscation, and de-identification image generation. We design the differential private k-anonymity algorithm that combines exponential differential privacy mechanism to introduce additional randomness to the average attributes of k nearest neighbors in random subset. The method we propose can achieve pleasant visual perception and balance the tradeoff between privacy and utility by adjustable parameters. Experiments demonstrate that our method is effective in identity protection and utility preservation.

References

1. T. Li, L. Lin, Anonymousnet: Natural face de-identification with measurable privacy, in *Proceedings of the IEEE/CVF Conference on Computer Vision and Pattern Recognition Workshops* (2019)
2. L. Sweeney, K-anonymity: A model for protecting privacy. Int. J. Uncertain. Fuzzin. Knowl.-Based Syst. **10**(5), 557–570 (2002). https://doi.org/10.1142/S0218488502001648
3. A. Machanavajjhala, D. Kifer, J. Gehrke, M. Venkitasubramaniam, l-diversity: privacy beyond k-anonymity. ACM Trans. Knowl. Discovery from Data **1**(1), 3–es (2007)
4. N. Li, T. Li, S. Venkatasubramanian, t-closeness: Privacy beyond k-anonymity and l-diversity, in *2007 IEEE 23rd International Conference on Data Engineering* (IEEE, Piscataway, 2007), pp. 106–115
5. H. Hukkelås, R. Mester, F. Lindseth, Deepprivacy: A generative adversarial network for face anonymization, in *International Symposium on Visual Computing* (Springer, Berlin, 2019), pp. 565–578
6. M. Maximov, I. Elezi, L. Leal-Taixé, Ciagan: Conditional identity anonymization generative adversarial networks," in *Proceedings of the IEEE/CVF Conference on Computer Vision and Pattern Recognition* (2020), pp. 5447–5456

7. R. Shokri, M. Stronati, C. Song, V. Shmatikov, Membership inference attacks against machine learning models, in *2017 IEEE Symposium on Security and Privacy (SP)* (2017), pp. 3–18.
8. B. Hitaj, G. Ateniese, F. Perez-Cruz, Deep models under the gan: Information leakage from collaborative deep learning, in *Proceedings of the 2017 ACM SIGSAC Conference on Computer and Communications Security* (2017), pp. 603–618
9. M. Fredrikson, S. Jha, T. Ristenpart, Model inversion attacks that exploit confidence information and basic countermeasures, in *Proceedings of the 22nd ACM SIGSAC Conference on Computer and Communications Security*, ser. CCS '15 (Association for Computing Machinery, New York, 2015), pp. 1322–1333. [Online]. Available: https://doi.org/10.1145/2810103.2813677
10. A.D. Sarwate, K. Chaudhuri, Signal processing and machine learning with differential privacy: algorithms and challenges for continuous data. IEEE Signal Process. Mag. **30**(5), 86–94, 2013
11. N. Papernot, M. Abadi, U. Erlingsson, I. Goodfellow, K. Talwar, Semi-supervised knowledge transfer for deep learning from private training data, in *ICLR* (2017)
12. Y. Zhu, X. Yu, M. Chandraker, Y.-X. Wang, Private-knn: Practical differential privacy for computer vision, in *Proceedings of the IEEE/CVF Conference on Computer Vision and Pattern Recognition (CVPR)* (2020), pp. 11851–11859
13. I.J. Goodfellow, J. Pouget-Abadie, M. Mirza, B. Xu, D. Warde-Farley, S. Ozair, A. Courville, Y. Bengio, Generative adversarial networks, in *Proceedings of the 27th International Conference on Neural Information Processing Systems*, vol. 2 (2014), pp. 2672–2680
14. X. Huang, Y. Li, O. Poursaeed, J. Hopcroft, S. Belongie, Stacked generative adversarial networks, in *Proceedings of the IEEE Conference on Computer Vision and Pattern Recognition* (2017), pp. 5077–5086
15. T. Kim, M. Cha, H. Kim, J.K. Lee, J. Kim, Learning to discover cross-domain relations with generative adversarial networks, in *International Conference on Machine Learning*, PMLR (2017), pp. 1857–1865
16. W. Shen, R. Liu, Learning residual images for face attribute manipulation, in *Proceedings of the IEEE Conference on Computer Vision and Pattern Recognition* (2017), pp. 4030–4038
17. P. Isola, J.-Y. Zhu, T. Zhou, A.A. Efros, Image-to-image translation with conditional adversarial networks, in *Proceedings of the IEEE Conference on Computer Vision and Pattern Recognition* (2017), pp. 1125–1134
18. G. Perarnau, J. van de Weijer, B. Raducanu, J.M. Álvarez, Invertible conditional gans for image editing (2016). arXiv e-prints arXiv–1611
19. Z. He, W. Zuo, M. Kan, S. Shan, X. Chen, Attgan: Facial attribute editing by only changing what you want. IEEE Trans. Image Process. **28**(11), 5464–5478 (2019)
20. Y. Choi, M. Choi, M. Kim, J.-W. Ha, S. Kim, J. Choo, Stargan: Unified generative adversarial networks for multi-domain image-to-image translation," in *Proceedings of the IEEE Conference on Computer Vision and Pattern Recognition* (2018), pp. 8789–8797
21. M. Liu, Y. Ding, M. Xia, X. Liu, E. Ding, W. Zuo, S. Wen, Stgan: A unified selective transfer network for arbitrary image attribute editing, in *2019 IEEE/CVF Conference on Computer Vision and Pattern Recognition (CVPR)* (2019), pp. 3668–3677
22. X. Li, S. Zhang, J. Hu, L. Cao, X. Hong, X. Mao, F. Huang, Y. Wu, R. Ji, Image-to-image translation via hierarchical style disentanglement (2021). arXiv preprint arXiv:2103.01456.
23. C. Dwork, Differential privacy: A survey of results, in *International Conference on Theory and Applications of Models of Computation*. ser. TAMC'08 (Springer, Berlin, 2008), pp. 1–19
24. B. Balle, G. Barthe, M. Gaboardi, Privacy amplification by subsampling: Tight analyses via couplings and divergences, in *Proceedings of the 32nd International Conference on Neural Information Processing Systems* (Curran Associates, Glasgow, 2018), pp. 6280–6290
25. I. Gulrajani, F. Ahmed, M. Arjovsky, V. Dumoulin, A. Courville, Improved training of wasserstein gans, in *Proceedings of the 31st International Conference on Neural Information Processing Systems* (2017)
26. Z. Liu, P. Luo, X. Wang, X. Tang, Deep learning face attributes in the wild, in *Proceedings of the IEEE International Conference on Computer Vision* (2015), pp. 3730–3738

27. W.L. Croft, J.-R. Sack, W. Shi, Obfuscation of images via differential privacy: From facial images to general images. Peer-to-Peer Netw. Appl. **14**, 1–29 (2021)
28. R. Zhang, P. Isola, A.A. Efros, E. Shechtman, O. Wang, The unreasonable effectiveness of deep features as a perceptual metric, in *Proceedings of the IEEE Conference on Computer Vision and Pattern Recognition* (2018), pp. 586–595
29. M. Heusel, H. Ramsauer, T. Unterthiner, B. Nessler, S. Hochreiter, Gans trained by a two time-scale update rule converge to a local nash equilibrium, in *Proceedings of the 31st International Conference on Neural Information Processing Systems* (2017), pp. 6629–6640
30. R. Zhang, P. Isola, A.A. Efros, E. Shechtman, O. Wang, The unreasonable effectiveness of deep features as a perceptual metric, in *Proceedings of the IEEE Conference on Computer Vision and Pattern Recognition (CVPR)* (2018)

Chapter 5
Differential Private Identification Protection for Face Images

5.1 Introduction

Today's popularity of smartphones allows people to take their face photos conveniently. Particularly, the blooming development of media and network techniques makes a vast amount of photos more approachable. At the same time, however, advanced image retrieval and face verification models allow to index and examine privacy relevant information more reliably than ever. Consequently, among those image sources exposed to the public with or without our awareness, the wide range of private information inadvertently leaked is severely underestimated [1].

Opportunities for misuse of the unprotected face image and advanced computer vision technologies are numerous and potentially disastrous [2]. Restrictive laws and regulations such as the General Data Protection Regulations (GDPR) [3] have taken effect. GDPR requires regular consent from the individual for any use of their personal data to guarantee data privacy; however, it also makes the creation of high quality datasets that include people becoming extremely challenging. Fortunately, if the data does not allow us to identify the corresponding individual, entities are free to use the data without consent. what is more, many computer vision tasks in practice, such as detection, tracking, or people counting, do not need to identify the people, but to detect them.

All the troubles and dilemmas mentioned above can be summarized to one issue: Given a face image, how can we create another image with similar appearance and the same background, while the real identity is hidden and face detectors are still allowed to work. Traditional anonymization techniques are mainly obfuscation-based and always significantly alter the original face. Other previous work in this field is sparse and limited in both practicality and efficacy: k-Same algorithm-

Main contents of this chapter have been published in "Wen, Y., Liu, B., Ding, M., Xie, R., & Song, L. (2022). IdentityDP: Differential private identification protection for face images. Neurocomputing, 501, 197–211."

Y. Wen et al., *Face De-identification: Safeguarding Identities in the Digital Era*,
https://doi.org/10.1007/978-3-031-58222-6_5

Fig. 5.1 IdentityDP for face anonymization. In each pair, left is the original image and right is the corresponding de-identified result. The results show that face identities are changed in a perceptually natural manner, and in the meantime, each pair of images still shares most of the information irrelevant to identity

based methods [4–8] fail to make full use of existing data and deliver fairly poor visual quality, adversarial perturbation-based methods [9–14] usually depend highly on the accessibility of the target system and require special training, and recent Generative Adversarial Network (GAN)-based methods [15–24] have trouble generating visually similar de-identified faces as well. Note that there exists a tradeoff between privacy protection and dataset utility [25, 26], and previous methods are unable to balance this matter.

To tackle these challenges, we propose IdentityDP, a framework designed to anonymize face images without significantly distorting the original images or destroying the availability of face detectors (see Fig. 5.1). Especially, individuals are allowed to have control over the anonymization procedure to get the most suitable results in practice. IdentityDP achieves this by helping users adding well-designed obfuscation to photos' high-level identity representations. For example, a user who wants to share photos on social media or the public web can add adjustable perturbations according to his demands through our framework before uploading them. The uploaded photos will look similar to the original ones, but when an adversary employs a general face verificator to compare the user's face images with the altered ones, it will indicate that they are from different people.

The proposed IdentityDP framework consists of three stages. Stage-I aims to perform facial representation disentanglement. We train a specially designed GAN for disentanglement between high-level identity representation and multilevel attribute representations in the feature space. Here the identity representation affects face verification systems to judge whether it is the same person, and the attribute representation guarantees the visual similarity. Stage-II carries out an ϵ-IdentityDP mechanism, where adjustable DP [27] perturbations are applied to the identity representation. Stage-III implements the image reconstruction. In more detail, we fix the well-trained GAN network in Stage-I and generate de-identified face images utilizing the perturbed identity representation as well as the original attribute representations. IdentityDP leverages both the GAN's outstanding ability to disentangle images' representations in the latent space and DP theory, managing to balance the tradeoff between image quality and privacy protection according to practical needs. In addition, our framework requires neither pre-annotation nor pre-detection of faces but can generate numerous anonymous results.

Our contributions in this work are as follows:

- We propose a general framework that is suitable for the de-identification of people in face images.
- As far as we know, we are the first to introduce the rigorously formulated DP theory into the face-anonymous task. The users are able to get not only high quality anonymous images but also an adjustable privacy protection mechanism.
- We demonstrate that our method does not require special training or targeted adjustments for many unauthorized identity verification systems or face datasets that have never seen before.
- We show that images anonymized by our method can be detected by common face detection models, so the processed images are still usable for identity-agnostic computer vision tasks (such as monitoring and tracking).
- We show that our de-identified method is significantly less computationally complex and consumes a small amount of computing resources.

The remainder of this chapter is organized as follows. In Section II, we summarize related work. Section III formalizes the face de-identification problem, introduces relevant DP theory, and proposes our assumptions. Section IV outlines the three-stage IdentityDP framework. Results of experiments analyzing the proposed IdentityDP method and comparisons with existing methods are reported in Section V. We conclude in Section VI with discussions of future research direction.

5.2 Related Work

In Chap. 3 of this book, we have classified face image de-identification methods according to the main technical means used. We introduce the current research status of corresponding methods category by category. Since in this chapter we propose a theoretically guaranteed face image identity protection scheme, here we specifically summarize the face image de-identification methods which can provide theoretical guarantees. These methods can be divided into the following three categories based on the theory that provides support.

5.2.1 Face De-identification Methods Guaranteed by k-Anonymity Theory

The k-anonymity theory was first proposed by Sweeney et al. in 1998 [28]. It was later modified and expanded and finally established in 2002 [29]. It can provide privacy protection for objects whose data-shape is (or can be regarded as) collections of quasi-identifiers. However, when encountering homogeneity attack [30], background knowledge attack [30], composition attacks [31], and the presence of supplementary data, the k-anonymity theory will not be able to provide privacy guarantees that meet theoretical expectations.

This privacy theory is adaptively adjusted by Newton et al. [4] based on the data form of face images and is summarized as the k-Same algorithm in the study of face de-identification. Before the rise of deep learning, k-anonymity theory was the most popular theoretical pillar when developing new guaranteed face identity privacy protection algorithms. In this book, the face image de-identification method based on the k-Same algorithm has been specifically introduced in Sect. 3.1.2.

Although this series of methods used to be the mainstay in the field of face de-identification, there are two significant shortcomings in them that caused them to be quickly iterated after the development of DNN. These two shortcomings are as follows: (1) The strict preconditions in the k-Same algorithm are difficult to meet in practical use, so the actual privacy performance generally cannot reach the theoretical guarantee value, and (2) the de-identified face generated by the k-Same algorithm often has lower visual quality than the original face and may result in double shadow effect and artifacts due to the dislocation between multiple faces. In particular, the existence of at least k-similar generated faces is not conducive to the normal use of these faces, so the utility performance of this kind of method is generally unsatisfactory.

5.2.2 Face De-identification Methods Guaranteed by t-Closeness Theory

The t-closeness theory is proposed by Li et al. in 2006 [32]. It requires that the distribution of a sensitive attribute in any equivalence class is close to the distribution of the attribute in the overall table (i.e., the distance between the two distributions should be no more than a threshold t). This theory effectively limits the amount of personal-specific information that an attacker can learn and can cope with attacks of globally distributed information that possesses sensitive characteristics, thus making up for the shortcomings of the above k-anonymity theory.

As far as we know, at present only the AnonymousNet [15] is a face de-identification method whose privacy protection is guaranteed by this t-closeness theory. AnonymousNet is an attribute manipulation-based method, which is described in detail in Sect. 3.1.4.1. It has four stages, of which the second stage implements the most critical identity privacy protection, which is accomplished by blurring facial attributes with PPAS algorithm that complies with the t-closeness theory. Although it can provide theoretically guaranteed privacy protection for identities, the faces de-identified by this method are significantly different from the original faces. In addition, this method is inconsistent with common sense, that is, it is generally believed that facial attributes (such as hairstyle, skin color, eyebrow thickness, etc.) can be changed through makeup, and these do not mean a change in identity. Therefore, even if this method leads to experimentally verified identity changes through cumulative changes of facial attributes, it is inconsistent with people's common sense and experience.

5.2.3 Face De-identification Method Guaranteed by Differential Privacy Theory

Face de-identification methods that provide privacy guarantees based on k-anonymity theory and t-closeness theory have shortcomings in both privacy and utility performance, so researchers begin to seek better privacy guarantees. With the development of DNN-based face representation learning and the progress in the field of database privacy protection, DP theory, as a rigorous privacy theory with practical protection capabilities, has begun to be selected as a theoretical guarantee by some methods. These methods treat the images as databases, then define different database records and add various DP noise on them, and ultimately achieve DP theory guaranteed privacy protection.

Some methods treat global pixels within an image as database records and design DP mechanisms for them. Fan proposes a pixelation method, DP-Pix [33], guaranteed by DP theory. DP-Pix first pixelates the source image and then designs a Laplace DP mechanism for the pixelated image. Although effectively protecting identities, this approach significantly reduces the utility of de-identified face images. Later, the same author proposes a blur-based method, DP-Blur [34], whose privacy protection is also guaranteed by the DP theory. The difference is that DP-Blur first pixelizes the input image and adds Gaussian DP perturbation. Then the image is upsampled back to the original size, and Gaussian blur is implemented. This approach also significantly damages the utility of the image. In order to improve the utility performance of generated images, the author proposes the DP-SVD method [35], which uses Singular Value Decomposition (SVD) to maintain the perceptual similarity between images. The concrete step is to first convert the face area into a feature vector and then perform random sampling that satisfies the metric privacy to achieve guaranteed privacy protection. The sampled feature vectors are inversely transformed to generate an anonymized image at last. However, the visual quality of the generated images is still unsatisfactory. Saleem et al. propose an interactive face image de-identification framework [36] including DP-Pix and DP-SVD methods. In addition, the privacy protection method guaranteed by DP theory proposed by Liu et al. [37] is achieved by adding global noise. Specifically, they first use image segmentation technology to transform the image gray matrix into a one-dimensional ordered data stream and then use a sliding window model to model the data stream. By comparing the similarity of data in adjacent sliding windows, they dynamically allocate the privacy budget and add Laplace noise.

Other methods treat the latent features of images as database records and design DP mechanisms for them. For example, Croft et al. [38] apply DP theory to protect the digital representation extracted from pixel intensities by a generative model and generate an output that looks like a real face through an AAM or a GAN model. In the context of preserving demographic information in images, this method can achieve comparable utility to the k-Same family methods. However, the quality of the generated images and the similarity to the original images are still not satisfactory. Later, Chen et al. introduce perceptual indistinguishability (PI) based on metric privacy [39] as a formal privacy notion particularly for images, which is a variant of DP theory that takes into account the perceptual similarity of face images, and propose PI-Net [40], an encoder–decoder architecture for image obfuscation with PI guarantee. In particular, they inject PI guaranteed noise into latent codes derived from GAN inversion and use triplet loss to cluster faces with similar facial attributes. These designs enable PI-Net to generate de-identified images that look realistic and satisfy user-defined facial attributes. However, since real images have been proven to be unable to faithfully invert back to the latent space of GAN [41], the process of image reconstruction based on GAN inverted features by PI-Net will cause the background of the generated image to change, and the face looks different from the original image. These do not meet the user's expectation.

There are two recent research works [42, 43] trying to implement the DP mechanism in images, both with the main new idea of injecting DP noise in the entire latent space. The biggest disadvantage of these two works is that the quality of the output is very sensitive to the size of the noise. Even a large privacy budget, that is, a small noise perturbation, will distort the generated face image. In addition, the work [44], a contemporaneous work with the study in this chapter, also achieves identity protection by applying the differential privacy mechanism to image embedding in the latent space obtained through GAN network. The difference is that those authors control the application of noise by using principal component analysis (PCA), so as to achieve a favorable tradeoff between privacy and utility. However, after protecting the identity privacy of the image, this method can only maintain two attributes as the original image, i.e., the same head pose and gender. In contrast, the method proposed in this chapter can keep the generated image more similar to the original image, and the overall visual effect is better.

5.3 Preliminaries

5.3.1 Problem Formulation

A face de-identification model can be viewed as a transformation function δ that maps a given face image X to a de-identified image $\hat{X}$, aiming to mislead face verification systems. Essentially, we are generating a new fake identity out of the input image. The problem can be formulated as follows:

$$\begin{gathered}\delta(X) = \hat{X} \\ s.t.: \text{Identity}\{X\} \neq \text{Identity}\{\hat{X}\}.\end{gathered} \tag{5.1}$$

Meanwhile, considering the utility of de-identified image, $\hat{X}$ should look similar to X as much as possible and be detectable by general face detectors.

5.3.2 Differential Privacy Theory

5.3.2.1 Differential Privacy

Differential privacy (DP) [27], a cryptography-inspired privacy-preserving model, guarantees that the likelihood of seeing an output on a given original dataset is close to the likelihood of seeing the same output on another dataset that differs from the original one in any single row. Here, the output could be another dataset, a statistical summary table, or a simple answer to a query, etc. Generally speaking, the basic idea of a DP mechanism is to introduce randomness into the original dataset, so that any individuals' information cannot be inferred by an adversary looking at the released output.

A formal definition of DP is shown below:

Definition 5.1 (ϵ-DP [45]) A randomized mechanism $\mathcal{T}$ gives ϵ-differential privacy if for any neighboring datasets D and D' differing on one element and all transcripts t,

$$\left| \ln \left(\frac{Pr[\mathcal{T}(D) = t]}{Pr[\mathcal{T}(D') = t]} \right) \right| \leq \epsilon. \tag{5.2}$$

This parameter ϵ, which is usually referred to as a privacy budget, is a bound on the ratio of the likelihood probabilities of seeing the same output on neighboring datasets. The smaller the value of ϵ, the stronger the privacy guarantee.

A random perturbation can be added to achieve the DP. Sensitivity calibrates the amount of noise for a specified query f of dataset D. Δf is the l_1-norm sensitivity defined as follows:

Definition 5.2 (l_1-Norm Sensitivity [45]) For any query f: $D \rightarrow \mathbb{R}$, l_1-norm sensitivity is the maximum l_1-norm of $f(D) - f(D')$, i.e.,

$$\Delta f = \max_{D,D'} ||f(D) - f(D')||_1. \tag{5.3}$$

The Laplace mechanism is one of the most generic mechanisms to guarantee differential privacy [46].

Definition 5.3 (Laplace Mechanism [45]) Given a function f: $D \rightarrow \mathbb{R}$, the following mechanism $\mathcal{T}$ provides the ϵ-differential privacy:

$$\mathcal{T}(D) = f(D) + Lap\left(\frac{\Delta f}{\epsilon}\right). \tag{5.4}$$

5.3.2.2 Local Differential Privacy

In traditional DP setting, there is a trusted curator who applies carefully calibrated random noise to the real values returned for a particular query. However, in many practical scenarios, the curator might not be trustworthy. In the local setting, there is no trusted third party, and the data needs to be randomized without the global knowledge. Local differential privacy (LDP) [47–49] is applicable to this case. It is considered to be a strong and rigorous notion of privacy that provides plausible deniability and is deemed to be a SOTA approach for privacy-preserving data collection and distribution.

Definition 5.4 (ϵ-LDP [50]) A randomized mechanism $\mathcal{A}$ satisfies ϵ-LDP, if for any two inputs v, v' and the set of all possible outputs $y \in \mathcal{Y}$, $\mathcal{Y}$ = $Range(\mathcal{A})$, $\mathcal{A}$ satisfies

$$Pr[\mathcal{A}(v) = y] \leq e^{\epsilon} \cdot Pr[\mathcal{A}(v') = y]. \tag{5.5}$$

And the sensitivity in this case equals

$$\Delta f = \max_{v,v' \in V} ||f(v) - f(v')||_1. \tag{5.6}$$

5.3.2.3 Two Important Properties

Our approach relies on two key properties of DP. The first is the widely used parallel composition property when designing mechanisms:

Property 5.1 (Parallel Composition [51]) Suppose we have a set of privacy mechanisms M = $\{M_1,\ldots,M_m\}$, and if each M_i provides ϵ_i privacy guarantee on a disjointed subset of the entire dataset, M will provide (max$\{\epsilon_1,\ldots,\epsilon_m\}$)-differential privacy.

The second is the well-known post-processing property:

Property 5.2 (Post-processing Property [52]) Any computation applied to the output of an (ϵ,δ)-DP algorithm remains (ϵ,δ)-DP.

For example, averaging, rounding, or any change to the output will not impact the privacy of the data. This means that an analyst can conduct any data post-processing on a released DP dataset and cannot reduce its privacy guarantee.

5.3.3 Face Verification and Our Assumptions

The key idea of face verification is to develop effective representations in feature space for reducing intra-personal variations while enlarging inter-personal differences [53]. The most ideal state is directly learning a mapping from face images to a compact feature space where distances precisely correspond to a measure of identity similarity. There are currently two main types of solutions: One is metric learning-based, and contrastive loss [54], center loss [55], and triplet loss [56] are proposed to enhance the discrimination power of features, and the other is angular margin-based, and many efforts [57–60] about angle margin penalty have greatly improved the verification accuracy. To some extent, anonymization can be considered as a task to protect someone's identity representations from being correctly classified.

Here we have an assumption that identity representations of one person in different feature spaces are interrelated. Once a face image's high-level representation in one feature space is disturbed into the wrong identity category, its identity representations in other feature spaces would also be classified incorrectly.

5.3.4 The Proposed IdentityDP Framework

For a given original clean face image X, our proposed IdentityDP framework can be used to generate its anonymous face images $\hat{X}$ in a controllable manner. We factor the face de-identification task into three stages. In the first stage, we use a person's image as input and disentangle the latent space information into two main representations, namely identity and attribute. Among them, identity representation is modeled by embedding features through an encoder, while attribute representations are modeled by multilevel embedding features through a decoder, and then the original face image is restored in an adaptively manner. In the second stage, we impose ϵ-IdentityDP perturbations on identity representation according to practical demands. In the third stage, we freeze all the parameters of the network

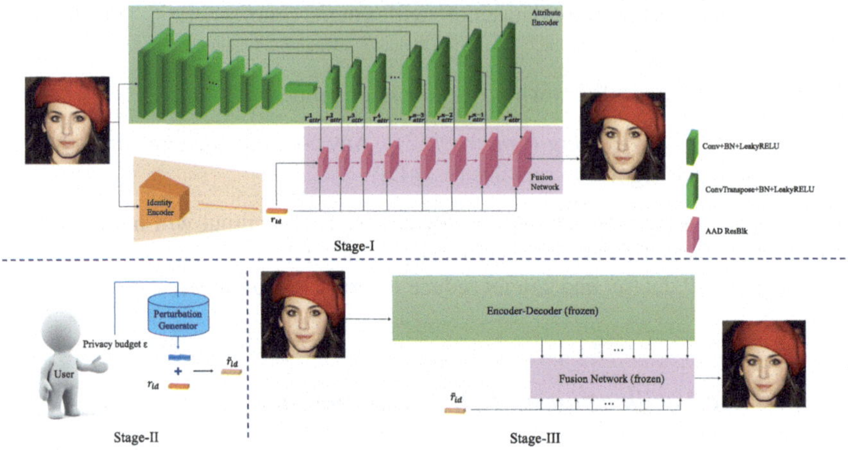

Fig. 5.2 Architecture of the proposed three-stage IdentityDP framework, which is based on a data-driven DNN and a Laplace ϵ-IdentityDP mechanism. Stage-I: training a network which can first extract disentangled high-level identity representation with multilevel attribute representations and then restore the original face; Stage-II: generating the perturbed identity representation under the ϵ-IdentityDP mechanism; and Stage-III: crafting anonymous faces from perturbed identity representation and original attribute representations through the frozen network

and synthesize anonymous face image with the perturbed identity representation. The overall architecture of the IdentityDP framework is shown in Fig. 5.2.

5.3.5 Stage-I: Facial Representations Disentanglement

In Stage-I, given an input face image, our goal is to represent the image using two disentangled representations, r_{id} and r_{att}. r_{id} is expected to contain all the information relevant to the identity, and r_{att} contains the rest of information carried by the image. We investigate how to generate satisfactory face images with a specific disentanglement intention (i.e., identity and attribute) in mind. The key idea is to explicitly guide the generation process by an appropriate representation of that intention. Therefore, our network consists of three components: (1) Identity Encoder, (2) Attribute Encoder, and (3) Fusion Generator.

Identity Encoder As mentioned before, studies on face verification and recognition have made arduous efforts in finding suitable face features that can reduce intra-personal variations while enlarging inter-personal differences, which is in line with our requirement of identity representation. Therefore, we choose a pretrained SOTA face recognition model [60] as our identity encoder, so as to exploit the existing technology to extract high-level identity representation in latent space. This pretrained model [60] can provide highly discriminative features for face recognition and has a clear geometric interpretation due to the exact correspondence to the geodesic distance on the hypersphere. The identity representation $r_{id}(X)$ is defined to be the last feature vector before the final FC layer, which can present off-the-shelf precise facial identity features to avoid training from scratch. It is denoted as

$$r_{id}(X) = f(X). \tag{5.7}$$

Attribute Encoder Attribute representation, which determines pose, expression, illumination, background, and so on, intuitively carries more spatial information than identity. Johnson et al. [61] have illustrated that low-level features tend to preserve image content and overall spatial structure, while high-level features tend to preserve color, texture, and exact shape. In order to preserve different level details, we employ multilevel feature maps to represent the attributes. In specific, we feed the input image X into a U-Net-like structure and then use the feature maps generated from the U-Net decoder as the attribute representations. More formally, we denote

$$r_{att}(X) = g(X) = \left\{ r_{att}^{1}(X), r_{att}^{2}(X), \ldots, r_{att}^{n}(X) \right\}, \tag{5.8}$$

where $r_{att}^{k}(X)$ represents the k-th level feature map from the U-Net decoder, and n is the number of feature levels.

This attribute encoder does not require any artificial annotations, and it extracts the attributes using self-supervised training: We require that the generated de-identified face $\hat{X}$ and the original face X have the same attributes embedding. The loss function will be introduced later in Eq. (5.16).

Fusion Network After obtaining the disentangled identity and attribute representations, we would like to learn a way to integrate them to reproduce the original face image, which will be used in our subsequent steps. Through a simple trial, we find that direct feature concatenation can easily lead to blurry results and is not expected to be used. Fortunately, Li et al. used *Adaptive Attentional Denormalization* (AAD) ResBlk [62] to achieve remarkable feature integration in multiple feature levels. They argued that the attention mechanism with denormalizations can make the effective regions of features more adaptive to adjust. This is an appealing property for fusion network since identity and attribute representations can participate in synthesizing different parts of the face. We integrate n AAD ResBlks to the body of our fusion network. As illustrated in Fig. 5.2, in Stage-I, after extracting the identity representation r_{id} and encoding multilevel attribute feature maps r_{att}, the fusion generator integrates them through cascaded AAD ResBlks to restore the original face image X:

$$X = h(r_{id}, r_{att}). \tag{5.9}$$

The training of $h(.)$ will be discussed in the following sections.

5.3.6 Stage-II: ϵ-IdentityDP Perturbation

Stage-II generates the perturbed identity representation under a novel Laplace ϵ-IdentityDP mechanism, which is defined as follows:

Definition 5.5 (ϵ-IdentityDP Mechanism) A randomized mechanism $\mathcal{M}$ satisfies ϵ-IdentityDP, i.e., if for any two inputs face images X, X' and the set of all possible outputs $y \in \mathcal{Y}$, $\mathcal{M}$ satisfies $Pr[\mathcal{M}(X) \in \mathcal{Y}] \leq e^{\epsilon} \cdot Pr[\mathcal{M}(X') \in \mathcal{Y}]$. For a face image X, if

$$f(X) = r_{id}(X) \tag{5.10}$$

and

$$\mathcal{M}(X) = r_{id}(X) + Lap\left(\frac{\Delta f}{\epsilon}\right) = \tilde{r}_{id}(X), \tag{5.11}$$

we say that $\mathcal{M}(X)$ satisfies ϵ-IdentityDP.

And the sensitivity is calculated as follows:

$$\Delta f = \max_{X,X'} ||r_{id}(X) - r_{id}(X')||_1. \tag{5.12}$$

To generate the perturbed identity representation in Eq. (5.11) and achieve the ϵ-IdentityDP mechanism, we employ a noise generator to generate suitable Laplace noise whose size equals the high-level identity representation according to specific privacy budget ϵ. To be more specific, in Stage-II, we employ a noise generator to generate Laplace noise whose size equals to $r_{id}(X)$ based on the selected privacy budget ϵ, and then we directly add the noise on $r_{id}(X)$ to obtain a perturbed identity representation $\tilde{r}_{id}$.

5.3.7 Stage-III: Image Reconstruction

Stage-III is conditioned on the obfuscated identity representation $\tilde{r}_{id}(X)$ and the original multilevel attribute features $r_{att}(X)$. In order to achieve good de-identified results, we freeze all the parameters of the well-trained fusion network in Stage-I and generate anonymous face image $\hat{X}$ by combining the perturbed identity representation $\tilde{r}_{id}$ and the original attribute representations $r_{att}(X)$ through the fusion network, which is formulated as

$$\hat{X} = h(\mathcal{M}(X), g(X)) = h(\tilde{r}_{\mathrm{id}}, r_{att}). \tag{5.13}$$

It can be approved that the generated image $\hat{X}$ follows ϵ-IdeneityDP.

Proof First, according to definition in Eq. (5.11), $\mathcal{M}(\mathcal{X})$ satisfies ϵ-IdentityDP:

$$\frac{Pr(\tilde{r}_{\text{id}}|f(X))}{Pr(\tilde{r}_{\text{id}}|f(X'))} = \prod_{i=1}^{m} \frac{\exp(-|r_{id(i)} - f(X)_i|/\frac{\Delta f}{\epsilon})}{\exp(-|r_{id(i)} - f(X')_i|/\frac{\Delta f}{\epsilon})}$$

$$= \prod_{i=1}^{m} \exp\left(\frac{\epsilon(|r_{id(i)} - f(X')_i| - |r_{id(i)} - f(X)_i|)}{\Delta f}\right)$$

$$\leq \prod_{i=1}^{m} \exp\left(\frac{\epsilon|f(X)_i - f(X')_i|}{\Delta f}\right)$$

$$= \exp\left(\frac{\epsilon \cdot \sum_{i=1}^{m} |f(X)_i - f(X')_i|}{\Delta f}\right)$$

$$= \exp\left(\frac{\epsilon \cdot \left\|f(X) - f(X')\right\|_1}{\Delta f}\right)$$

$$\leq \exp(\epsilon),$$

where the first inequality follows from that $|a| - |b| \leq |a - b|$ for any $a, b \in \mathbb{R}$. The rest of proof follows from the post-processing property of DP. Hence, we can conclude that if the identity representation is treated with DP noises, then the reconstructed face image $\hat{X}$ also satisfies the ϵ-IdentityDP defined in Definition 5.5.

5.3.8 Training Process

In Stage-I, we need to build a network which can not only disentangle identity and attribute representations but also restore the original input face image from these two representations.

We utilize adversarial training for this framework. Let L_{adv} be the adversarial loss to make $\hat{X}$ realistic. It is implemented as a multi-scale discriminator [63] on the downsampled output images:

$$L_{adv}(\hat{X}, X) = \log D_{img}(X) + \log(1 - D_{img}(\hat{X})). \tag{5.14}$$

An identity preservation loss is used to preserve the identity of the source. It is formulated as

$$L_{id} = 1 - \cos(r_{id}(\hat{X}), r_{id}(X)), \tag{5.15}$$

where $\cos(\cdot, \cdot)$ represents the cosine similarity of two vectors. We also use the attribute preservation loss, which is defined as half of the sum of the squared Euclidean distances between the multilevel attribute representations from X and

$\hat{X}$. More formally, we denote

$$L_{att} = \frac{1}{2}\sum_{k=1}^{n} \left\| r_{att}^{k}(\hat{X}) - r_{att}^{k}(X) \right\|_2^2. \tag{5.16}$$

Besides, we set pixel-level $\mathcal{L}_2$ distance as the reconstruction loss to guarantee the visual similarity, which is formulated as

$$L_{rec} = \frac{1}{2}\left\| \hat{X} - X \right\|_2^2. \tag{5.17}$$

The full objective to train our network in the first stage is a weighted sum of above losses as

$$L_{total} = L_{adv} + \lambda_{att} L_{att} + \lambda_{id} L_{id} + \lambda_{rec} L_{rec}, \tag{5.18}$$

where λ_{att}, λ_{id}, and λ_{rec} are the weight parameters for balancing different terms.

In practice, GAN is hard to train, so adjusting the training strategy according to real-time generation effect is necessary. In order to use visualization tools to judge our training effect and make appropriate adjustments in time, we extract identity and attribute representations from two faces randomly sampled from the training dataset and then fuse them together during the training process. It is worth noting the reconstruction loss should be set to $L_{rec} = 0$ when the two faces are different.

5.3.9 Some Discussions About Our Research Topic

(1) The motivation of using DP for face de-identification
The reason we need to perform de-identification is that the face image is a personal identifier which should be properly protected from the privacy perspective. In more detail, we want to prevent the information leakage of personal identities from releasing face images, and we hope that the privacy protection level can be measured by a formal criterion. Meanwhile, although DP is the most widely used notion for privacy protection, there is no effective and formal DP definition or mechanism in the context of image. This motivates us to use DP to prevent identity information leakage from face images, and we propose the IdentityDP method which makes an initial contribution to this meaningful research topic.

(2) Are we just doing adversarial attack-based privacy protection?
Initially, an adversarial attack is perceived as an "attack" method to mislead AI models, i.e., adding small (often human invisible) perturbation to the input data sample so as to corrupt the prediction of a deep learning model. Although there have been a few recent studies [14, 64] that explored the idea of adversarial

attacks for privacy protection, these methods differ significantly from our proposed method in the following two aspects:

- Adversarial attack-based privacy protection methods usually assume a machinery adversary, e.g., a deep learning model from previous work. As the adversarial perturbation is often small, the protection provided is not necessarily effective against human eyes. In contrast, our proposed method considers both human and machine as adversaries and provides effective privacy protection against both types of adversaries.
- There is no formal and strict privacy guarantee provided by the adversarial attack-based privacy protection methods, while the privacy level of our proposed IdentityDP is clearly defined and rigorously guaranteed by the DP criterion.

(3) Are we just doing differentially private machine learning?
Researchers in the field of differentially private deep learning [65] are work on preventing model itself from releasing private information of its training datasets and maintaining a manageable cost in software complexity, training efficiency, and model quality at the same time. However, it is different from our research topic of face de-identification. De-identification is a process which aims to remove all identification information of the person from an image or video while maintaining as much information on the action and its context with a similar looking appearance [23, 66]. Our concentration is to protect the private identity information of face images, but not to prevent our model from releasing private information of our training face datasets. In more detail, the role of machine learning in these two tasks is different: Their topic is to make machine learning system private, i.e., machine learning system is the target of privacy protection; however, our topic is to use machine learning techniques to enhance privacy protection (i.e., prevent the information leakage of personal identifiers from released face images). Therefore, these are two different research topics.

(4) Recent researches on DP-based face de-identification
Applying DP in images is a promising research topic because of the increasing concerns on image privacy, especially face privacy, and there are a few recent work [42, 43] to study this problem. They all try to implement DP into images, but in different ways. The main idea of these methods is to inject DP noise in the whole feature (latent) space. The disadvantage is that the photo's quality is very sensitive to the amount of noise, and even a small noise perturbation (large epsilon value) will make the photo distorted. Our work solves this problem by only adding noise to the disentangled identity representation. The essential point of our proposed method is that the noise needed for de-identification is much smaller than the existing methods, as the disentangled identity vector has a much smaller norm than the whole latent space vector. In addition, Laplace mechanism is the most often used mechanism to achieve a strict DP privacy guarantee. While other mechanisms such as Gaussian mechanism and Exponential mechanism may also be used, they are not as popular as Laplace mechanism. Hence, the existing methods [42, 43] that implement DP for images

all adopt the Laplace mechanism, and we select the Laplace DP mechanism in our method at the second stage too.

5.4 Experiments

5.4.1 Experimental Setup

(1) *Datasets*: We choose the CelebA-HQ dataset, which contains 30K high-resolution celebrity images with diverse demographic information like age, gender, and race [67], to train our network in Stage-I. We randomly select 27K images for training and 3K for testing. Moreover, in order to demonstrate our generalization ability and compare with conditional comparisons conveniently, we also test IdentityDP on the CelebA [68] datasets. All images are aligned and cropped to size 256×256 covering the whole face, as well as some background regions.
(2) *Comparison methods*: To validate the effectiveness of the proposed IdentityDP framework, we compare it to traditional anonymization methods as well as SOTA methods.

- Traditional anonymization methods. We use Pixelization, Noise, and Blur of faces.
- State-of-the-art methods. We select four methods: AnonymousNet [15], DeepPrivacy [19], CIAGAN [24], and Fawkes [14].

5.4.2 Evaluation Metrics

We evaluate all methods in privacy metrics as well as utility metrics.

(1) Privacy metrics. Two different metrics are used to measure the performance of privacy protection.

   - *Identity Distance* $\mathcal{ID\_DIS}$. We employ FaceNet identification model [56] based on Inception-ResNet backbone, pretrained on two public datasets: CASIA-WebFace [69] and VGGFace2 [70], whose LFW accuracy can reach 99.05 and 99.65% individually. The output distance of FaceNet can indicate the pairs of input faces' identity difference.
   - *Protection success rate* $\mathcal{PSR}$. Besides publicly available datasets and known model architectures for academic usage, we also wish to understand the performance of IdentityDP on public facial verification services that people may touch in daily life. Therefore, Microsoft Azure Face [71] is employed to evaluate real-world effectiveness of a method. It gives judgement of whether

the input pairs are of the same people. The protection success rate is the proportion of faces that are judged as different from the original ones.

(2) Utility metrics. Three different metrics are used to evaluate the utility of processed images.

- *PSNR* and *SSIM*. We choose peak-signal-to-noise ratio (PSNR) as well as structural similarity index measure (SSIM) as two objective measures of similarity between anonymous results and original faces.
- *Face detection rate* $\mathcal{FDR}$. We evaluate whether the processed images are still usable for subsequent identity-agnostic computer vision tasks by performing face detection using HOG [72] detector, and we calculate the proportion of faces that can be detected in the protected images.

5.4.3 Implementation Details

We implement our framework as shown in Fig. 5.2. The number of attribute representations is set to $n = 8$ (Eq. (5.8)). The detailed network structure is given in Fig. 5.3. Specifically, we build the network almost according to the description in literature [62], except that we select the pretrained SOTA face recognition model, ArcFace [60], whose output is a 512-dimensional vector as our identity encoder, while the dimension of identity vector in the original FaceShifter is 256. Besides, we also modify the corresponding channels in Fusion Network that are related to the high-level identity representation to adapt to this change. In the training process, we use the Adam optimizer [73] with momentum parameters $\beta_1 = 0, \beta_2 = 0.999$. The learning rate is set to 0.0004. The parameters in Eq. (5.18) are set to $\lambda_{att} = \lambda_{rec} = 10, \lambda_{id} = 5$. The size of identity representation is $512 \times 1 \times 1$, and the upper bound of sensitivity is 512.

5.4.4 ϵ-IdentityDP Mechanism Analysis

To explicitly understand the DP mechanism in our proposed IdentityDP, we design an experiment to explore how the privacy budget ϵ affects the face anonymization performance. First of all, we extract every test image's identity representation and calculate the l_1-norm sensitivity Δf, i.e., $\Delta f = \max_{X,X'} ||r_{id}(X) - r_{id}(X')||_1$, $X, X' \in$ *test datasets*. Then we increase ϵ from 1 to 800 and accordingly adjust the IdentityDP framework. Since our ϵ-IdentityDP mechanism $\mathcal{M}(X)$ is $\mathcal{M}(X) = r_{id}(X) + Lap(\frac{\Delta f}{\epsilon})$, we double ϵ for better display effect, and 100 anonymous faces are generated for every test face under each ϵ. Besides, we technically explore the case when ϵ varies uniformly in the interval (0,1] where the perturbation changes

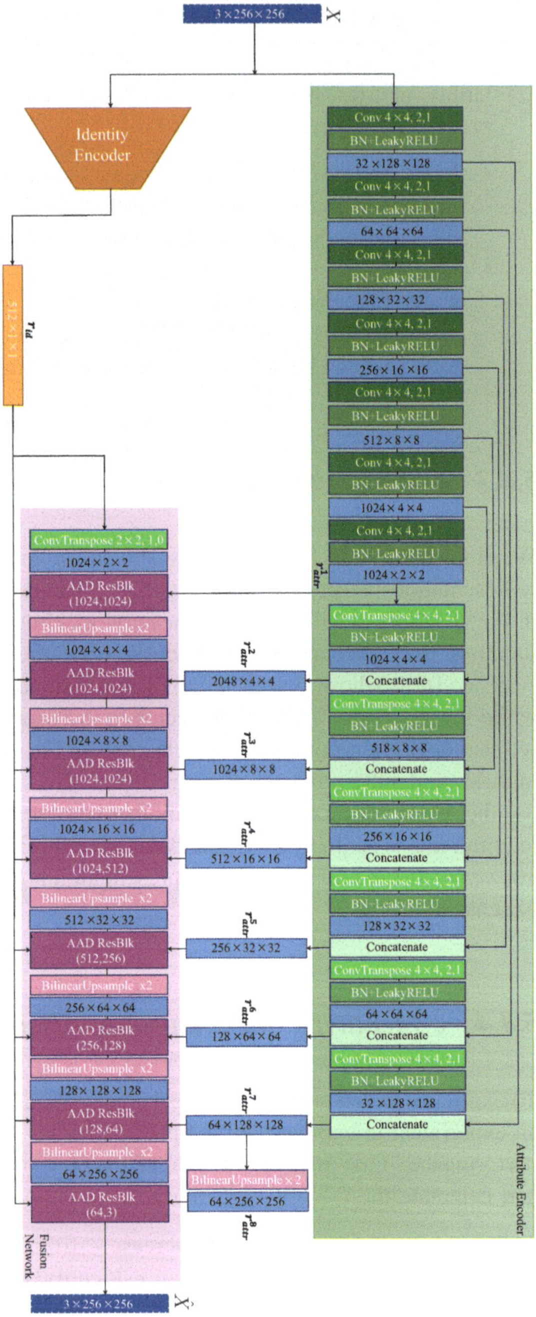

Fig. 5.3 Network structure of the proposed neural network in Stage-I. *Conv k,s,p* represents a Convolutional Layer with kernel size k, stride s, and padding p. *ConvTranspose k,s,p* represents a Transposed Convolutional Layer with kernel size k, stride s, and padding p. All *LeakyReLUs* have $\alpha = 0.1$. *AAD ResBlk* (c_{in}, c_{out}) represents an AAD ResBlk [62] with input and output channels of c_{in} and c_{out}

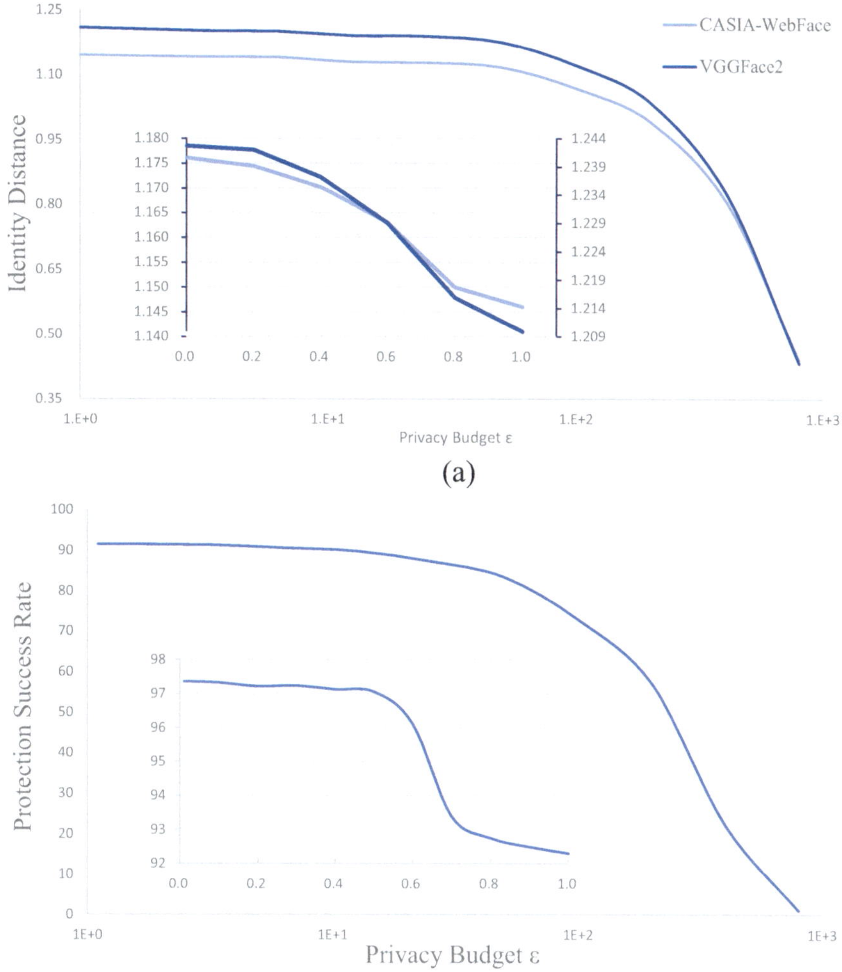

(b)

Fig. 5.4 Identity protection performance: (**a**) the identity distance calculated by FaceNet model trained on CASIAWebFace and VGGFace2 datasets respectively; (**b**) the Protection success rate calculated through public facial verification service [71]

more dramatically. Finally, various statistical mean metric values are calculated at each ϵ value.

For privacy protection, when ϵ increase from 0.01 to 800, Fig. 5.4a shows that the average identity distance decreases gradually, and Fig. 5.4b shows that the protection success rate decreases from 97.367 to 1.125%, illustrating that a smaller privacy budget guarantees better de-identified results. We show anonymous image

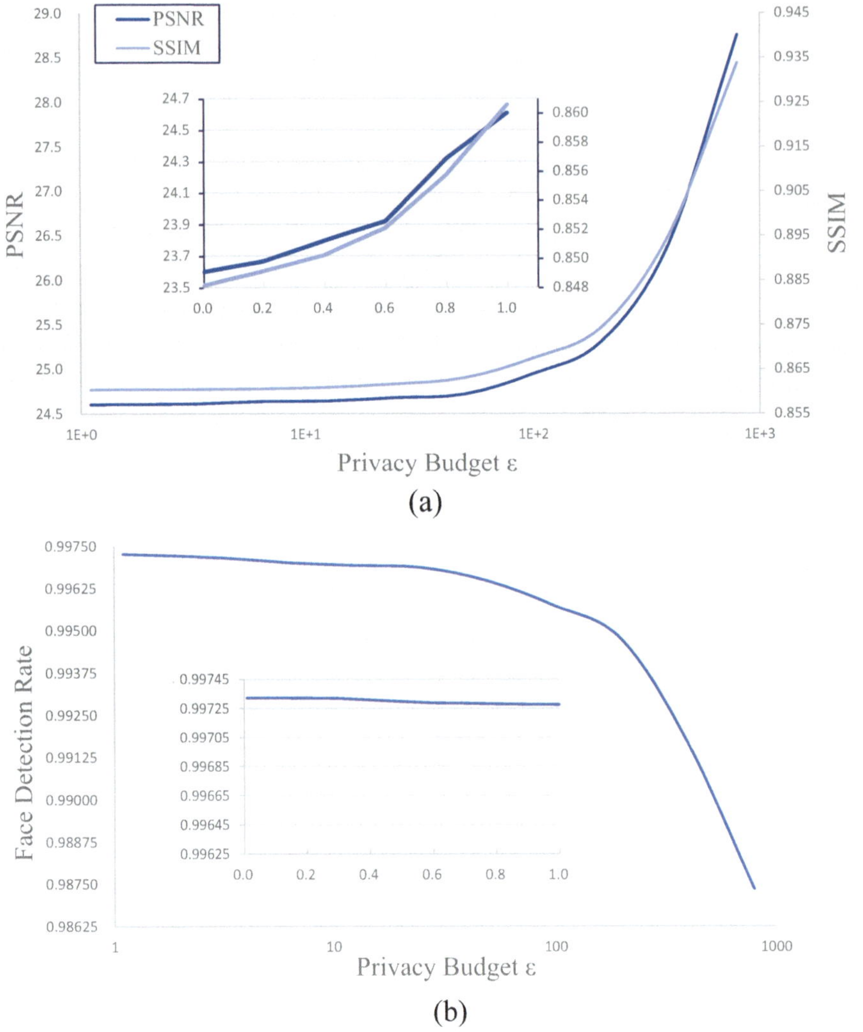

Fig. 5.5 Image utility performance: (**a**) PSNR and SSIM and (**b**) the Face detection rate calculated through HOG detector

whose identity distance is closest to the mean distance under every ϵ in Fig. 5.6, which also implies the diversity of our de-identified results.

For data utility, Fig. 5.5a plots PSNR and SSIM vs. ϵ, indicating that the visual similarity gets better as the privacy budget increases. Figure 5.5b shows that our face detection rate always remains at a high level, demonstrating that identity-agnostic computer vision technologies can still work on our processed faces. Specially, when the privacy budget ϵ is small (i.e., a strong privacy protection), the subtle differences between the de-identified faces and the corresponding original ones can be perceived

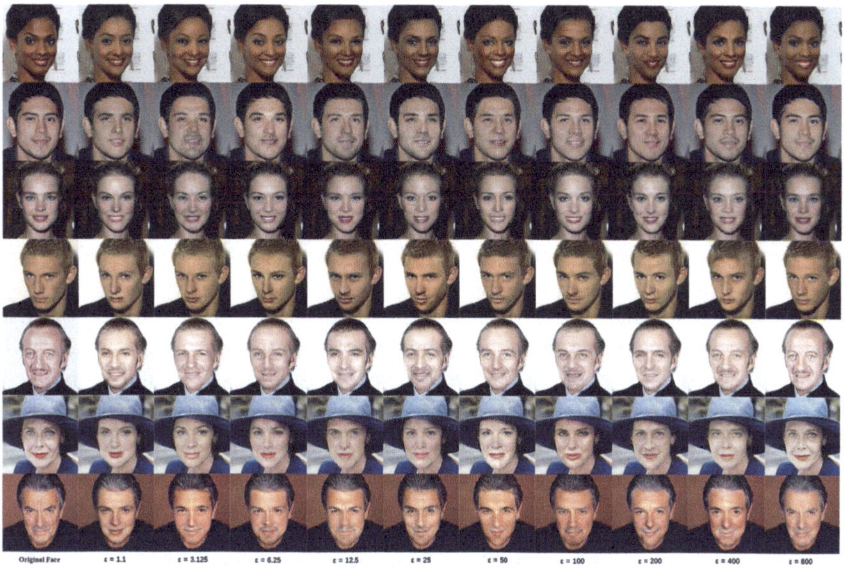

Fig. 5.6 Qualitative comparison of the influence of parameter ϵ. The first column shows the original face images. The rest columns demonstrate anonymous face whose identity distance is closest to the mean distance under every ϵ

by humans easily (e.g., different eyebrow shapes, different iris colors, and different lip shapes), while they still share a great visual similarity on the whole (Fig. 5.6).

Furthermore, an unexpected issue is that the face detection rate decreases slightly as ϵ increases. After research, we find the reason that partially severely blocked faces in test dataset can recover some facial features in the blocked area using our framework, resulting in the detection of originally undetectable faces.

Figure 5.1 illustrates some de-identified results in pairs, where left is the original image and right is the de-identified result generated by our framework. It demonstrates that human identities are obfuscated in a perceptually natural manner; in the meantime, each pair of images still shares similar appearance, as well as the same expression and background. It is worth noticing that our results can well retain the unique attributes of characters, such as rare hairstyles, beards, glasses, and other accessories, which is hard to achieve in previous GAN-based methods.

Based on a large number of experiments, we get some experience in choosing a suitable privacy budget value: If the image's hue is light or the people's expression is exaggerated, a smaller privacy budget should be chosen. In fact, we believe that setting the privacy budget to any positive number less than 10 can get an advanced anonymization result successfully, and we recommend the user to set their privacy budget between 0.5 and 7 to obtain anonymous face efficiently with quite well-preserved appearance.

5.4.5 Comparisons with Traditional Methods

In this subsection, the following traditional methods are implemented: (1) Pixelization: We cluster face region's pixels that are close in 2D space and color space and then replace each cluster (8×8, 16×16) with its average value. (2) Noise: We add Gaussian noise ($\sigma = 9, 49$) on each pixel's RGB value of the face region. (3) Blur: Following Ryoo et al. [74], we downsample the face region to extreme low resolution (7×7, 19×19) and then upsample back. We set the privacy budget to 6. It can be seen that for the fairness of comparison, we select two parameters for each traditional method: One aims to make the identity distance close to our approach; at this time, the utility metrics are mainly compared, and the other aims to make PSNR or SSIM close to our method; at this time, the privacy metrics are mainly compared.

Figure 5.7 shows the qualitative results. It is obvious that our approach achieves a great advantage in visual similarity as well as realism. The detailed quantitative results are shown in Table 5.1. Although PSNR and SSIM of our method are lower

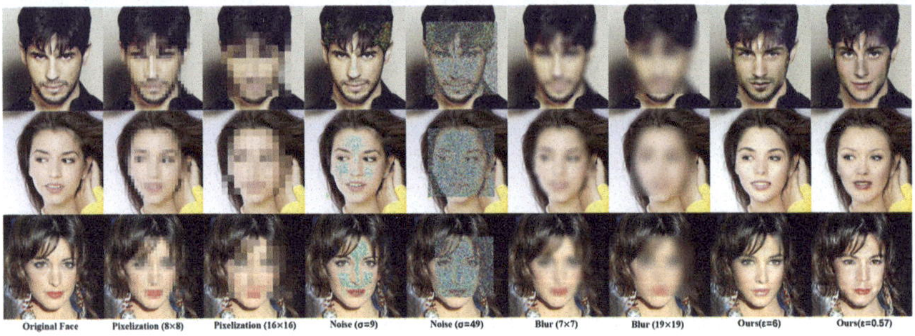

Fig. 5.7 Qualitative comparison with traditional methods. From left to right: original faces, faces de-identified by Pixelization (4×4, 8×8), Noise ($\sigma = 9, 18$), Blur (8×8, 16×16), and faces generated by our method ($\epsilon = 6$ and $\epsilon = 0.57$)

Table 5.1 Quantitative evaluation on CelebA-HQ datasets under different metrics

	$\mathcal{ID\_DIS}$					
Method	CASIA	VGGFace2	$\mathcal{PSR}$	PSNR	SSIM	$\mathcal{FDR}$
Pixelization(8×8)	0.8646	0.8993	0	26.735	0.7671	0.923
Pixelization(16×16)	1.1541	1.2195	0.017	23.926	0.7223	0.058
Noise($\sigma = 9$)	0.3317	0.2723	0.002	23.831	0.8312	0.986
Noise($\sigma = 49$)	1.1267	1.0280	0.012	14.370	0.5533	0.425
Blur(7×7)	0.8491	0.8380	0	27.405	0.806	0.888
Blur(19×19)	1.1102	1.1857	0.669	24.829	0.7719	0.518
DeepPrivacy	1.0860	1.1829	0.961	21.012	0.7808	0.989
Fawkes	0.7267	0.8585	0	**35.898**	**0.9487**	0.985
Ours($\epsilon = 6$)	1.1403	1.2012	0.908	24.640	0.8606	0.997
Ours($\epsilon = 0.57$)	**1.1644**	**1.2307**	**0.967**	23.909	0.8519	**0.997**

than some traditional methods (i.e., Pixelization (8 × 8), Pixelization (16 × 16), Blur (7 × 7), and Blur (19 × 19)), we can see that (1) Pixelization (8 × 8) and Blur (7 × 7) get much smaller values than ours under the $\mathcal{ID\_DIS}$ metric and get zero under the $\mathcal{FDR}$ metric, which means that their de-identification effect is almost nil when faced with advanced face verification technology. (2) Pixelization (16 × 16) makes the value of $\mathcal{FDR}$ drop sharply to 0.058, which means that its de-identified faces can no longer be used in other identity-agnostic applications. This seriously damages the utility of face images. In addition, one of its privacy metrics, $\mathcal{PSR}$, is close to 0, which means that Pixelization (16 × 16) is vulnerable to public face verification API. (3) Except for a slightly better PSNR value, Blur (19 × 19) is inferior to our method under all other metrics, especially the $\mathcal{PSR}$ and $\mathcal{FDR}$. Therefore, our method can not only effectively de-identify faces but also maintain the results' utility. We perform best in preventing advanced face verification and maintaining the privacy-utility tradeoff.

5.4.6 Comparisons with SOTA Methods

In this subsection, we compare our IdentityDP with SOTA face de-identification methods. Among them, DeepPrivacy and Fawkes are trained and tested on CelebA-HQ datasets. AnonymousNet and CIAGAN require pre-annotations and are trained on CelebA datasets, so we transfer our framework on CelebA and compare with them for fairness. We evaluate performance with these methods, respectively.

(1) *Comparisons with Attribute Manipulation-Based Anonymization*: Facial attributes, including gender, age, haircut, and so on, should be an important reference for identifying faces' identities, especially affecting human's subjective judgment. Therefore, manipulating face attributes to make faces anonymous seems reasonable. AnonymousNet, a privacy-preserving attribute selection algorithm for facial image obfuscation, is a typical representative.
Figure 5.8 shows the anonymous faces generated from our framework and those from AnonymousNet. Due to the change of several face attributes, the anonymous face generated by AnonymousNet is often visually different from the original face, especially when modifying gender, which is not conducive to the normal use of the images. In contrast, our method achieves significant improvement in visual similarity. As can be seen from Table 5.2, our method performs better under both privacy metrics and utility metrics, not to mention that AnonymousNet requires detailed data annotations. Moreover, it is worth noticing that although anonymous faces generated by AnonymousNet are visually very different from the original one, face verification service API can still judge them correctly, which suggests that general face attributes are not directly related to human identity.
(2) *Comparisons with Conditional Inpainting-Based Anonymization*: Exposure of faces is the source of private information leakage. Therefore, some methods

Fig. 5.8 Qualitative comparison of our method with AnonymousNet [15] and CIAGAN [24]. The top row shows original faces, and the second row and the third row show corresponding anonymous faces generated by AnonymousNet and CIAGAN. The last two rows show our results (ϵ = 6 and ϵ = 0.57)

Table 5.2 Quantitative evaluation on CelebA datasets under different metrics

	$\mathcal{ID\_DIS}$					
Method	CASIA	VGGFace2	$\mathcal{PSR}$	PSNR	SSIM	$\mathcal{FDR}$
AnonymousNet	0.8896	1.0589	0.295	18.892	0.7192	0.892
CIAGAN	0.8155	1.0271	0.945	21.863	0.7401	0.958
Ours($\epsilon = 6$)	0.9345	1.0918	0.905	**23.353**	**0.8188**	0.986
Ours($\epsilon = 0.57$)	**0.9622**	**1.1176**	**0.961**	22.7639	0.8005	**0.987**

directly feed their networks with face-removing images as well as auxiliary annotations to automatically generate anonymous human faces. In this way, the generator never touches original faces, which ensures the removal of any privacy-sensitive information. DeepPrivacy is such a method which requires two annotations: A bounding box to identify the privacy-sensitive area and a sparse seven keypoints pose estimation of the face. It generates de-identified faces considering the original pose and image background. We compare our method with it.
Figure 5.9 reports the difference of methods. We can see that the face generated by DeepPrivacy can maintain the facial pose well but is not visually similar to the original image. Besides, distortions and artifacts often occur. Our method produces more visual-pleasing anonymous faces which look similar to the original one. Table 5.1 shows quantitative results, and when ϵ is set to 6, our method is slightly inferior to DeepPrivacy in terms of privacy protection but has remarkable data utility improvement; moreover, when ϵ is set to 0.57, all metrics of our method outperform DeepPrivacy.

(3) *Comparisons with Conditional ID-Swapping-Based Anonymization*: Since anonymizing a face is intended to hide its original identity, swapping the original ID with others becomes a straightforward idea. Conditioned on face landmark and masked background image of the input image, CIAGAN generates a new fake identity out of the input image to achieve anonymization. We compare images generated from our proposed framework and those from CIAGAN. From Fig. 5.8 we can see that the two methods produce comparable results, while our method enjoys a better visual similarity and less artifacts. Table 5.2 shows the quantitative results. When ϵ is set to 6, CIAGAN protects privacy better from the perspective of $\mathcal{PSR}$; however, when setting ϵ to 0.57, we outperform CIAGAN under all metrics and maintain a better visual similarity on the whole.
Moreover, CIAGAN has some notable flaws: (1) It needs to borrow someone else's identity as operation guidance, which may affect the privacy and security of the identity provider, (2) Faces de-identified by CIAGAN are visually similar to original ones only when the fake ID provider shares the same gender, a similar age, as well as similar makeup with the original ID owner, which makes it not very convenient to use in practice, (3) CIAGAN cannot maintain certain special attributes, such as glasses, heavy makeup, and thick beard, unless the identity provider also has, and (4) CIAGAN depends on landmark detection to provide pre-annotations, which tends to miss the face that has not been detected in the anonymization process. In contrast, our approach does not have these problems, since our method does not need the assistance of other identities, can retain the special attributes of original faces, and does not need pre-annotations. In summary, our method surpasses CIAGAN in privacy protection while maintaining a more similar appearance to their original ones, and our proposed method significantly performs better in terms of the utility metrics.

(4) *Comparisons with Adversarial Perturbation-Based Anonymization*: De-identified methods based on adversarial examples are continuously popular

Fig. 5.9 Qualitative comparison of our method with DeepPrivacy [19] and Fawkes [14]. The top row shows original faces, and the second row and the third row show corresponding anonymous faces generated by DeepPrivacy and Fawkes. The last two rows show our results (ϵ = 6 and ϵ = 0.57)

because their anonymous results are almost the same as the original images. However, their performance depends largely on the accessibility of the target system's internal parameters or special training on the target system. Fawkes, as one of the latest representatives, is selected as our comparison.

Figure 5.9 demonstrates that Fawkes can generate faces that look extremely like the original one, except for a few strange spots that sometimes appear. We just provide a comparable result. However, Table 5.1 shows that Fawkes gets much smaller values than ours under the $\mathcal{ID\_DIS}$ metric and gets zero under the $\mathcal{FDR}$ metric, which suggests that its de-identification effect is almost nil when faced with advanced face verification technology. In contrast, although our method suffers from less visual similarity, it works significantly better in

Table 5.3 Additional quantitative evaluation with SOTA methods on LFW datasets

Method	FaceNet model		FID
	CASIA	VGGFace2	
Original	0.965 ± 0.016	0.986 ± 0.010	0
AnonymousNet	0.037 ± 0.015	0.044 ± 0.016	6.8479
DeepPrivacy	0.029 ± 0.012	0.039 ± 0.014	2.7122
CIAGAN	0.019 ± 0.008	0.034 ± 0.015	2.1756
Fawkes	0.898 ± 0.010	0.917 ± 0.012	**1.2681**
Ours($\epsilon = 6$)	0.019 ± 0.010	0.031 ± 0.015	2.0201
Ours($\epsilon = 0.57$)	$\mathbf{0.016 \pm 0.011}$	$\mathbf{0.024 \pm 0.014}$	2.0437

preserving face privacy, which is the most important for the de-identification task.

(5) *Additional Discussion*: To make the comparison more convincing and fairer, we follow the evaluation protocol that has been used in CIAGAN and add two experiments with the SOTA methods to evaluate the performance of privacy and utility, respectively.

Firstly, we use the evaluation method for privacy protection, which is conducted on the LFW benchmark. In this experiment, we employ two FaceNet identification models (pretrained on CASIA-WebFace [69] and VGGFace2 [70]), and the main evaluation metric is the true acceptance rate. Table 5.3 presents the results on de-identified LFW image pairs for a given person, while the de-identification method is applied to the second image of each pair. It can be seen that all the SOTA methods can let the true positive rate drop from almost 0.99 to less than 0.05 except Fawkes. In particular, when ϵ is 0.57, our method yields the lowest true positive rate when two FaceNet models pretrained on CASIA dataset and VGGFace2 dataset are employed.

Then we evaluate the utility of the images by using the FID score on LFW dataset, as it can measure the distance between the generated distribution and the real distribution. The results are shown in Table 5.3. Among the methods that can effectively drop the true acceptance rate and well protect the identity information, our method achieves the best FID score. It demonstrates that our de-identified images exhibit a closer similarity to the original ones in terms of data distribution, which is consistent with our intuitive expectation.

Besides, there is no formal and strict privacy guarantee provided by the SOTA privacy protection methods, while the privacy level of our proposed IdentityDP is clearly defined and rigorously guaranteed by the DP criterion. Therefore, our method has the advantage of providing provable and strict privacy guarantee.

To better evaluate the visual perception of the results generated by our method, we also conduct a user study to assess the quality of identity anonymization for human observers of the proposed model. Given an original image, two de-identified images are generated using the two recommended privacy budgets (6 and 0.57) to form two image pairs in total. For each image pair, the participants are asked to evaluate whether they are the same person. We provide four options, namely

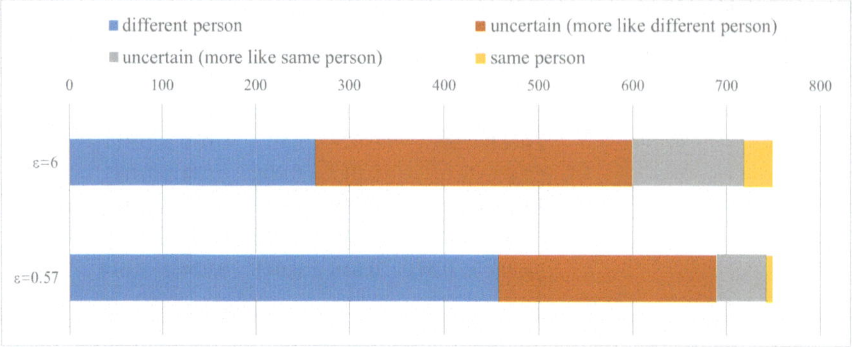

Fig. 5.10 Results of user study under two recommended privacy budget ϵ values

Table 5.4 Face detection rate on CelebA-HQ and CelebA datasets by using MTCNN

Dataset	CelebA-HQ				CelebA			
Method	[19]	[14]	Ours(ϵ=6)	Ours(ϵ=0.57)	[15]	[24]	Ours(ϵ=6)	Ours(ϵ=0.57)
$\mathcal{FDR}$	0.999	**1**	**1**	**1**	0.978	0.995	0.997	**0.999**

"different person," "uncertain (more likely different person)," "uncertain (more likely the same person)," and "same person." We present 100 image pairs from 50 original images to 15 human subjects and collect results. One thousand five hundred votes are collected, and the number of votes under two recommended privacy budgets is shown in the form of bar chart. The results in Fig. 5.10 demonstrate that most users can perceive the change in identity, which indicates that the de-identification performance of our method is good. Moreover, using the smaller recommended privacy budget can better ensure face anonymization for human observers.

To be more comprehensive, we add an experiment that calculates the face detection rate $\mathcal{FDR}$ by using a popular advanced detector, MTCNN [75]. Table 5.4 presents the results. It can be seen that there is an overall increase in $\mathcal{FDR}$, and our method achieves the best subsequent utility guarantee, i.e., the de-identified images can still be used in subsequent identity-agnostic computer vision applications.

5.4.7 Generalization Ability

Our IdentityDP provides great generalization to various face images. In previous experiments, it has been proved by showing remarkable qualitative and quantitative results on CelebA, a dataset that our IdentityDP has never trained on before. To further demonstrate the robustness of our method, we apply our framework to face images from the very difficult inputs of [76]. As can be seen in Fig. 5.11, our method is robust to very challenging illuminations.

Fig. 5.11 Our de-identification results on examples labeled as challenging or very challenging in the NIST Face Recognition Challenge [76]. The first row shows original faces, and the following row shows our corresponding de-identified results

Fig. 5.12 Our anonymization results on challenging artistic portraits. The first and the third row show the artistic portraits, while the second and the fourth row show our corresponding anonymous results

In addition, we apply our framework on artistic portraits. All artworks are taken from Wikiart.org. Figure 5.12 shows the interesting results, illustrating that faces in different styles are anonymized successfully without causing significant distortions or artifacts.

5.4.8 *Computational Overhead*

In this subsection, we evaluate our computational overheads for anonymizing faces. IdentityDP adds little overhead for processing, as the only additions are a random noise tensor. On an NVIDIA GTX 1080 Ti, IdentityDP takes on average 0.329s per image. The low computational overhead is beneficial to process a large amount of face images.

5.5 Conclusion and Future Work

In this chapter, we propose the IdentityDP framework that combines DP mechanisms with DNNs to achieve image privacy protection for the first time. Our framework consists of three stages: deep representations disentanglement, ϵ-IdentityDP perturbation, and image reconstruction. In our framework, DP perturbation is directly added on to the identity representation to ensure privacy protection, while the attribute representation is unchanged and it preserves visual similarity well. Furthermore, the adjustable privacy budget guarantees the diversity of anonymization results. Experiments demonstrate the effectiveness of our framework in terms of privacy protection and image utility and produce satisfactory results compared with the traditional and SOTA methods. Moreover, our framework has a good generalization ability. In the future, we will further explore the tradeoff between user privacy and authorized use of work. In addition, extending this work to videos and achieving temporal consistency would be an interesting direction.

References

1. P. Liu, Y. Xu, Q. Jiang, Y. Tang, Y. Guo, L.E. Wang, X. Li, Local differential privacy for social network publishing. Neurocomputing **391**, 273–279 (2020)
2. Y. Li, Y. Wang, D. Li, Privacy-preserving lightweight face recognition. Neurocomputing **363**, 212–222 (2019)
3. E. Commission, *2018 Reform of EU Data Protection Rules* (2018)
4. E.M. Newton, L. Sweeney, B. Malin, Preserving privacy by de-identifying face images. IEEE Trans. Knowl. Data Eng. **17**(2), 232–243 (2005)
5. R. Gross, E. Airoldi, B. Malin, L. Sweeney, Integrating utility into face de-identification, in *International Workshop on Privacy Enhancing Technologies* (Springer, Berlin, 2005), pp. 227–242
6. R. Gross, L. Sweeney, F. De la Torre, S. Baker, Model-based face de-identification, in *2006 Conference on Computer Vision and Pattern Recognition Workshop (CVPRW'06)* (IEEE, Piscataway, 2006), pp. 161–161
7. L. Du, M. Yi, E. Blasch, H. Ling, Garp-face: Balancing privacy protection and utility preservation in face de-identification, in *IEEE International Joint Conference on Biometrics* (IEEE, Piscataway, 2014), pp. 1–8

8. A. Jourabloo, X. Yin, X. Liu, Attribute preserved face de-identification, in *2015 International Conference on Biometrics (ICB)* (IEEE, Piscataway, 2015), pp. 278–285
9. S. Komkov, A. Petiushko, AdvHat: Real-world adversarial attack on ArcFace face ID system (2019). arXiv preprint arXiv:1908.08705
10. S.J. Oh, M. Fritz, B. Schiele, Adversarial image perturbation for privacy protection a game theory perspective, in *2017 IEEE International Conference on Computer Vision (ICCV)* (IEEE, Piscataway, 2017), pp. 1491–1500
11. A. Shafahi, W.R. Huang, M. Najibi, O. Suciu, C. Studer, T. Dumitras, T. Goldstein, Poison frogs! targeted clean-label poisoning attacks on neural networks, in *Advances in Neural Information Processing Systems* (2018), pp. 6103–6113
12. B. Liu, J. Xiong, Y. Wu, M. Ding, C.M. Wu, Protecting multimedia privacy from both humans and AI, in *2019 IEEE International Symposium on Broadband Multimedia Systems and Broadcasting (BMSB)* (IEEE, Piscataway, 2019), pp. 1–6
13. C. Zhu, W.R. Huang, A. Shafahi, H. Li, G. Taylor, C. Studer, T. Goldstein, Transferable clean-label poisoning attacks on deep neural nets (2019). *arXiv preprint arXiv:1905.05897*
14. S. Shan, E. Wenger, J. Zhang, H. Li, H. Zheng, B.Y. Zhao, Fawkes: Protecting privacy against unauthorized deep learning models, in *29th {USENIX} Security Symposium ({USENIX} Security 20)* (2020), pp. 1589–1604
15. T. Li, L. Lin, AnonymousNet: Natural face de-identification with measurable privacy, in *Proceedings of the IEEE Conference on Computer Vision and Pattern Recognition Workshops* (2019), pp. 0–0
16. H.-P. Wang, T. Orekondy, M. Fritz, InfoScrub: Towards attribute privacy by targeted obfuscation (2020). arXiv preprint arXiv:2005.10329
17. Q. Sun, L. Ma, S. Joon Oh, L. Van Gool, B. Schiele, M. Fritz, Natural and effective obfuscation by head inpainting, in *Proceedings of the IEEE Conference on Computer Vision and Pattern Recognition* (2018), pp. 5050–5059
18. Z. Ren, Y. Jae Lee, M.S. Ryoo, Learning to anonymize faces for privacy preserving action detection, in *Proceedings of the European Conference on Computer Vision (ECCV)* (2018), pp. 620–636
19. H. Hukkelås, R. Mester, F. Lindseth, DeepPrivacy: A generative adversarial network for face anonymization, in *International Symposium on Visual Computing* (Springer, Berlin, 2019), pp. 565–578
20. Y. Wu, F. Yang, Y. Xu, H. Ling, Privacy-protective-GAN for privacy preserving face de-identification. J. Comput. Sci. Technol. **34**(1), 47–60 (2019)
21. B. Meden, R.C. Mallı, S. Fabijan, H.K. Ekenel, V. Štruc, P. Peer, Face deidentification with generative deep neural networks. IET Signal Process. **11**(9), 1046–1054 (2017)
22. Q. Sun, A. Tewari, W. Xu, M. Fritz, C. Theobalt, B. Schiele, A hybrid model for identity obfuscation by face replacement, in *Proceedings of the European Conference on Computer Vision (ECCV)* (2018), pp. 553–569
23. O. Gafni, L. Wolf, Y. Taigman, Live face de-identification in video, in *Proceedings of the IEEE International Conference on Computer Vision* (2019), pp. 9378–9387
24. M. Maximov, I. Elezi, L. Leal-Taixé, CIAGAN: Conditional identity anonymization generative adversarial networks, in *Proceedings of the IEEE/CVF Conference on Computer Vision and Pattern Recognition* (2020), pp. 5447–5456
25. R. Hasan, E. Hassan, Y. Li, K. Caine, D.J. Crandall, R. Hoyle, A. Kapadia, Viewer experience of obscuring scene elements in photos to enhance privacy, in *Proceedings of the 2018 CHI Conference on Human Factors in Computing Systems* (2018), pp. 1–13
26. B. Rassouli, D. Gündüz, Optimal utility-privacy trade-off with total variation distance as a privacy measure. IEEE Trans. Inf. Forens. Secur. **15**, 594–603 (2019)
27. C. Dwork, Differential privacy: A survey of results, in *International Conference on Theory and Applications of Models of Computation* (Springer, Berlin, 2008), pp. 1–19
28. P. Samarati, L. Sweeney, Protecting privacy when disclosing information: k-anonymity and its enforcement through generalization and suppression. (1998). Technical Report. SRI International Computer Science Laboratory.

29. L. Sweeney, k-anonymity: a model for protecting privacy. Int. J. Uncert. Fuzzin. Knowl.-Based Syst. **10**(05), 557–570 (2002)
30. A. Machanavajjhala, D. Kifer, J. Gehrke, M. Venkitasubramaniam, l-diversity: privacy beyond k-anonymity. ACM Trans. Knowl. Discovery from Data (TKDD) **1**(1), 3–es (2007)
31. A. Basu, T. Nakamura, S. Hidano, S. Kiyomoto, k-anonymity: Risks and the reality, in *2015 IEEE Trustcom/BigDataSE/ISPA*, vol. 1 (IEEE, Piscataway, 2015), pp. 983–989
32. N. Li, T. Li, S. Venkatasubramanian, t-closeness: Privacy beyond k-anonymity and l-diversity, in *2007 IEEE 23rd International Conference on Data Engineering* (IEEE, Piscataway, 2006), pp. 106–115
33. L. Fan, Image pixelization with differential privacy, in *Data and Applications Security and Privacy XXXII: 32nd Annual IFIP WG 11.3 Conference, DBSec 2018, Bergamo, Italy, July 16–18, 2018, Proceedings 32* (Springer, Berlin, 2018), pp. 148–162
34. L. Fan, Differential privacy for image publication, in *Theory and Practice of Differential Privacy (TPDP) Workshop*, vol. 1, no. 2 (2019), p. 6
35. L. Fan, Practical image obfuscation with provable privacy, in *2019 IEEE International Conference on Multimedia and Expo (ICME)* (IEEE, Piscataway, 2019), pp. 784–789
36. M.U. Saleem, D. Reilly, L. Fan, DP-shield: Face obfuscation with differential privacy, in *Advances in Database Technology* (2022)
37. C. Liu, J. Yang, W. Zhao, Y. Zhang, J. Li, C. Mu, Face image publication based on differential privacy. Wirel. Commun. Mobile Comput. **2021**, 1–20 (2021)
38. W.L. Croft, J.-R. Sack, W. Shi, Differentially private obfuscation of facial images, in *Machine Learning and Knowledge Extraction: Third IFIP TC 5, TC 12, WG 8.4, WG 8.9, WG 12.9 International Cross-Domain Conference, CD-MAKE 2019, Canterbury, UK, August 26–29, 2019, Proceedings 3* (Springer, Berlin, 2019), pp. 229–249
39. K. Chatzikokolakis, M.E. Andrés, N.E. Bordenabe, C. Palamidessi, Broadening the scope of differential privacy using metrics, in *Privacy Enhancing Technologies: 13th International Symposium, PETS 2013, Bloomington, IN, July 10–12, 2013. Proceedings 13* (Springer, berlin, 2013), pp. 82–102
40. J.-W. Chen, L.-J. Chen, C.-M. Yu, C.-S. Lu, Perceptual indistinguishability-net (PI-Net): Facial image obfuscation with manipulable semantics, in *Proceedings of the IEEE/CVF Conference on Computer Vision and Pattern Recognition* (2021), pp. 6478–6487
41. K. Preechakul, N. Chatthee, S. Wizadwongsa, S. Suwajanakorn, Diffusion autoencoders: Toward a meaningful and decodable representation, in *Proceedings of the IEEE/CVF Conference on Computer Vision and Pattern Recognition* (2022), pp. 10619–10629
42. B. Liu, M. Ding, H. Xue, T. Zhu, D. Ye, L. Song, W. Zhou, DP-image: Differential privacy for image data in feature space (2021). arXiv preprint arXiv:2103.07073
43. T. Li, C. Clifton, Differentially private imaging via latent space manipulation (2021). arXiv preprint arXiv:2103.05472
44. W.L. Croft, J.-R. Sack, W. Shi, Differentially private facial obfuscation via generative adversarial networks. Future Gener. Comput. Syst. **129**, 358–379 (2022)
45. C. Dwork, F. McSherry, K. Nissim, A. Smith, Calibrating noise to sensitivity in private data analysis, in *Theory of Cryptography Conference* (Springer, Berlin, 2006), pp. 265–284
46. C. Dwork, A. Roth et al., The algorithmic foundations of differential privacy. Found. Trends Theor. Comput. Sci. **9**(3–4), 211–407 (2014)
47. S.P. Kasiviswanathan, H.K. Lee, K. Nissim, S. Raskhodnikova, A. Smith, What can we learn privately? SIAM J. Comput. **40**(3), 793–826 (2011)
48. J.C. Duchi, M.I. Jordan, M.J. Wainwright, Local privacy and statistical minimax rates, in *2013 IEEE 54th Annual Symposium on Foundations of Computer Science* (IEEE, Piscataway, 2013), pp. 429–438
49. P. Kairouz, S. Oh, P. Viswanath, Extremal mechanisms for local differential privacy. Adv. Neur. Inf. Process. Syst. **27**, 2879–2887 (2014)
50. T. Wang, J. Blocki, N. Li, S. Jha, Locally differentially private protocols for frequency estimation, in *26th {USENIX} Security Symposium ({USENIX} Security 17)* (2017), pp. 729–745

51. T. Zhu, G. Li, W. Zhou, S.Y. Philip, *Differential Privacy and Applications* (Springer, Berlin, 2017)
52. M. Bun, T. Steinke, Concentrated differential privacy: Simplifications, extensions, and lower bounds, in *Theory of Cryptography Conference* (Springer, Berlin, 2016), pp. 635–658
53. M. Wang, W. Deng, Deep face recognition: A survey. Neurocomputing **429**, 215–244 (2021)
54. Y. Sun, X. Wang, X. Tang, Sparsifying neural network connections for face recognition, in *Proceedings of the IEEE Conference on Computer Vision and Pattern Recognition* (2016), pp. 4856–4864
55. Y. Wen, K. Zhang, Z. Li, Y. Qiao, A discriminative feature learning approach for deep face recognition, in *European Conference on Computer Vision* (Springer, Berlin, 2016), pp. 499–515
56. F. Schroff, D. Kalenichenko, J. Philbin, FaceNet: A unified embedding for face recognition and clustering, in *Proceedings of the IEEE Conference on Computer Vision and pattern Recognition* (2015), pp. 815–823
57. W. Liu, Y. Wen, Z. Yu, M. Li, B. Raj, L. Song, SphereFace: Deep hypersphere embedding for face recognition, in *Proceedings of the IEEE Conference on Computer Vision and Pattern Recognition* (2017), pp. 212–220
58. F. Wang, J. Cheng, W. Liu, H. Liu, Additive margin softmax for face verification. IEEE Signal Process. Lett. **25**(7), 926–930 (2018)
59. H. Wang, Y. Wang, Z. Zhou, X. Ji, D. Gong, J. Zhou, Z. Li, W. Liu, CosFace: Large margin cosine loss for deep face recognition, in *Proceedings of the IEEE Conference on Computer Vision and Pattern Recognition* (2018), pp. 5265–5274
60. J. Deng, J. Guo, N. Xue, S. Zafeiriou, ArcFace: Additive angular margin loss for deep face recognition, in *Proceedings of the IEEE Conference on Computer Vision and Pattern Recognition* (2019), pp. 4690–4699
61. J. Johnson, A. Alahi, L. Fei-Fei, Perceptual losses for real-time style transfer and super-resolution, in *European Conference on Computer Vision* (Springer, Berlin, 2016), pp. 694–711
62. L. Li, J. Bao, H. Yang, D. Chen, F. Wen, FaceShifter: Towards high fidelity and occlusion aware face swapping (2019). arXiv preprint arXiv:1912.13457
63. T. Park, M.-Y. Liu, T.-C. Wang, J.-Y. Zhu, Semantic image synthesis with spatially-adaptive normalization, in *Proceedings of the IEEE Conference on Computer Vision and Pattern Recognition* (2019), pp. 2337–2346
64. H. Xue, B. Liu, M. Din, L. Song, T. Zhu, Hiding private information in images from AI, in *ICC 2020–2020 IEEE International Conference on Communications (ICC)*, Dublin, Ireland. (IEEE, Piscataway, 2020).
65. M. Abadi, A. Chu, I. Goodfellow, H.B. McMahan, I. Mironov, K. Talwar, L. Zhang, Deep learning with differential privacy, in *Proceedings of the 2016 ACM SIGSAC Conference on Computer and Communications Security* (2016), pp. 308–318
66. P. Agrawal, P. Narayanan, Person de-identification in videos. IEEE Trans. Circ. Syst. Video Technol. **21**(3), 299–310 (2011)
67. T. Karras, T. Aila, S. Laine, J. Lehtinen, Progressive growing of GANs for improved quality, stability, and variation (2017). arXiv preprint arXiv:1710.10196
68. Z. Liu, P. Luo, X. Wang, X. Tang, Deep learning face attributes in the wild, in *Proceedings of the IEEE International Conference on Computer Vision* (2015), pp. 3730–3738
69. D. Yi, Z. Lei, S. Liao, S.Z. Li, Learning face representation from scratch (2014). arXiv preprint arXiv:1411.7923
70. Q. Cao, L. Shen, W. Xie, O.M. Parkhi, A. Zisserman, VGGFace2: A dataset for recognising faces across pose and age, in *2018 13th IEEE International Conference on Automatic Face & Gesture Recognition (FG 2018)* (IEEE, Piscataway, 2018), pp. 67–74
71. Microsoft azure face API. https://azure.microsoft.com/en-us/services/cognitive-services/face/
72. N. Dalal, B. Triggs, Histograms of oriented gradients for human detection, in *2005 IEEE Computer Society Conference on Computer Vision and Pattern Recognition (CVPR'05)*, vol. 1 (IEEE, Piscataway, 2005), pp. 886–893

73. D.P. Kingma, J. Ba, Adam: A method for stochastic optimization (2014). arXiv preprint arXiv:1412.6980
74. M.S. Ryoo, B. Rothrock, C. Fleming, H.J. Yang, Privacy-preserving human activity recognition from extreme low resolution (2016). arXiv preprint arXiv:1604.03196
75. K. Zhang, Z. Zhang, Z. Li, Y. Qiao, Joint face detection and alignment using multitask cascaded convolutional networks. IEEE Signal Process. Lett. **23**(10), 1499–1503 (2016)
76. P.J. Phillips, J.R. Beveridge, B.A. Draper, G. Givens, A.J. O'Toole, D.S. Bolme, J. Dunlop, Y.M. Lui, H. Sahibzada, S. Weimer, *An Introduction to the Good, the Bad, & the Ugly Face Recognition Challenge Problem* (.IEEE, Piscataway, 2011)

Chapter 6
Personalized and Invertible Face De-identification

6.1 Introduction

Face images are generally considered to contain abundant private information. The earliest techniques obfuscated privacy-sensitive information by pixel-level processing which has been proved vulnerable and has poor effects on utility [1]. Recent GAN-based methods [2, 3] improve the quality and utility of de-identification results remarkably. What is more, the research on disentangled representations [4, 5] contributes to transforming the identity without changing the other facial attributes, so that the de-identified results look similar to the original.

However, most de-identification methods only focus on the protection phase. Considering that when we share pictures with close friends or in some specific scenarios like criminal investigations, it is hoped to use the original image instead of the de-identified. Therefore, how to restore the original image is also a critical task. Moreover, notice that the tradeoff between privacy and utility poses a major challenge for all privacy-preserving methods, and different levels of privacy are required in different scenarios. In summary, we believe that an ideal comprehensive de-identification method should (a) avoid deteriorating nonsensitive information like facial expression, behavior, and so on, (b) control the degree of privacy protection according to different application scenarios, and (c) be able to restore the original image under security conditions.

To achieve the above targets, this chapter proposes a personalized and invertible face de-identification method, in which the user can set a password and the privacy level for a given face image. The main framework can be summarized in the following three stages: (1) extract disentangled identity and attributes to make sure

---

Main contents of this chapter have been published in "Cao, J., Liu, B., Wen, Y., Xie, R., & Song, L. (2021). Personalized and invertible face deidentification by disentangled identity information manipulation. In Proceedings of the IEEE/CVF International Conference on Computer Vision (pp. 3334–3342)."

Y. Wen et al., *Face De-identification: Safeguarding Identities in the Digital Era*,
https://doi.org/10.1007/978-3-031-58222-6_6

the generated results share the same attributes and visual similarity with the input, (2) calculate the protected or restored identity with the identity modification module based on the password p and privacy level parameter d, and (3) implement the image reconstruction.

In summary, our main contributions are as follows:

- A general framework that can transform identity of the input while ensuring the other attributes keep similar.
- Personalized de-identification results can be generated with the user-specific password, and the degree of identity variation can be controlled.
- The original image can be restored if and only if the corresponding encryption password is provided.
- Experimental results show that compared with existing methods, our approach can generate de-identified results with better performance of both privacy and utility, in addition to better quality recovery results.

6.2 Problem Formulation

Our identity conversion algorithm mainly possesses de-identification $\mathcal{F}$ and restoration $\mathcal{F}^{-1}$, which both require the input of source face image X, the user-specific password p, and a privacy level parameter d. The password can determine the direction of identity variation and d can control the variation degree. Inspired by Gu et al. [6], we mathematically formulate our problem in this section.

De-identification In order to achieve the effectiveness of identity protection, we aim that the protected image will have different identity information from the original, which can be formulated as

$$I(\mathcal{F}(X, p, d)) \neq I(X), \tag{6.1}$$

where $\mathcal{F}(X, p, d)$ indicates the de-identified X with parameters p and d, $I(X)$ represents the identity of image X. Considering the utility of de-identified results, we hope that $\mathcal{F}(X, p, d)$ looks similar to X as well as the face region and keypoints can still be detected by the face detector.

Diversity We can set different passwords p to generate diverse de-identification results, which can promote the security of identity protection.

$$I(\mathcal{F}(X, p_1, d)) \neq I(\mathcal{F}(X, p_2, d)), p_1 \neq p_2. \tag{6.2}$$

Controllability We can control the similarity between the de-identified image and the original by the adjustable parameter d as

$$D(\mathcal{F}(X, p, d_1), X) > D(\mathcal{F}(X, p, d_2), X), d_1 > d_2, \tag{6.3}$$

where $D(X, Y)$ means the identity distance between image X and Y, and the larger distance indicates lower similarity.

Recoverability If the user takes the de-identified result $\mathcal{F}(X, p, d)$ and corresponding password p and d as input, the origin image X can be restored successfully, which can be formulated as

$$\mathcal{F}^{-1}(\mathcal{F}(X, p, d), p, d) = \hat{X}, \tag{6.4}$$

$$I(X) = I(\hat{X}). \tag{6.5}$$

However, if the attacker tries to restore the image without the right identity encryption password, he/she can only get the image with another identity instead of the original one.

$$\mathcal{F}^{-1}(\mathcal{F}(X, p_1, d), p_2, d) = \hat{Y}, p_1 \neq p_2, \tag{6.6}$$

$$I(X) \neq I(\hat{Y}). \tag{6.7}$$

In addition to the above, we also expect that both the de-identified and the restored have high distinct image quality and satisfactory visual perception.

6.3 Our Approach

The framework of training process is shown in Fig. 6.1 and that of protection process and recovery process is presented in Fig. 6.2, which mainly consists of two encoders E_{id} and E_{attr}, an identity modification module M, and a generator G. In the first

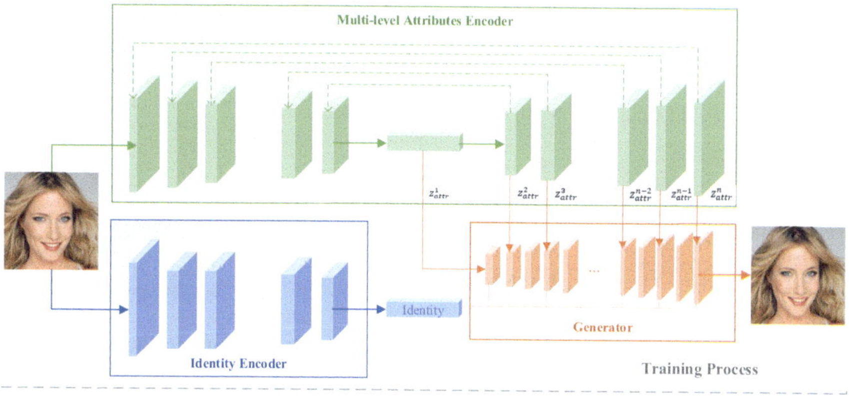

Fig. 6.1 The framework of training process, which includes the identity encoder, the multilevel attributes encoder, and the generator

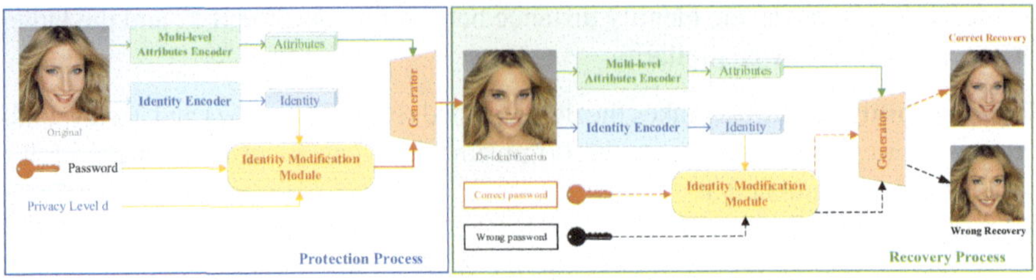

Fig. 6.2 The framework of protection process and recovery process

stage, we extract the image representations and disentangle them into identity z_{id} and attributes z_{attr}. Second, we calculate the protected identity z_{new} or the restored $\hat{z}_{id}$ by the identity modification module M. Finally, G generates de-identified results based on z_{new} and z_{attr} (or the restored based on $\hat{z}_{id}$ and z_{attr}). Each part will be described in detail in this section.

6.3.1 Network Architecture

6.3.1.1 Identity Encoder

Similar to most research on disentangled representations of identity and attributes, we use a pretrained face recognition model as the identity encoder E_{id}, and the identity representation z_{id} is taken from the last feature vector before the final fully connected layer.

6.3.1.2 Attribute Encoder

In order to retain better details of attributes like expression, pose, illumination, and so on, we design to represent the attribute representations as multilevel feature maps. We employ a U-Net-like structure and define the feature vectors obtained from each layer of the U-Net decoder as attributes embedding.

$$z_{attr} = \left\{ z_{attr}^{1}, z_{attr}^{2}, \cdots z_{attr}^{n} \right\}. \tag{6.8}$$

6.3.1.3 Identity Modification Module

The identity modification module mainly edits the identity embedding with latent space manipulation. Most of the state-of-the-art (SOTA) face recognition or face verification models such as ArcFace [7], CosFace [8], and SphereFace [9], convert identity features to the hyperspherical space and use cosine similarity based on angles. This has motivated us to conclude that rotating the identity vector is a more effective way to change identity information compared with other vector operations like translation. Additionally, considering the feasibility of restoration, we hope to introduce a definite modification process instead of introducing randomness like noise. Therefore, we realize de-identification process $z_{new} = M(z_{id}, p, d)$ or restoration process $\hat{z}_{id} = M^{-1}(z_{new}, p, d)$ by changing the phase of identity embedding. In more detail, during the protection process, we first perform normalization on z_{id} and extract a reference vector z_r from the pre-defined reference identity vector library. Each reference identity z_r is obtained by randomly selecting k different identities from the training set to combine and normalize, which ensures that there is no real corresponding identity to avoid identity leakage. Then the component vector z_{90} that is orthogonal to z_{id} in z_r will be decomposed and form a set of orthogonal bases with z_{id}. Through the combination of the basis vectors set, the new identity representation z_{new} after z_{id} rotation with the degree of θ on the hyperplane can be formulated as

$$z_{new} = z_{id} \cos\theta + z_{90} \sin\theta, \tag{6.9}$$

where z_{90} determines the direction of rotation and z_{new} may correspond to the identity of an unreal person. Since z_{90} is a component vector of z_r, we can introduce the mapping $z_r = f(p)$, $\theta = g(d)$ to control of the direction and degree of identity variation with p and d. In the recovery phase, we can calculate the original identity with the inverse operations, and more detailed calculations will be introduced in Sect. 6.3.4.

6.3.1.4 Generator

We are required to design a network to implement image reconstruction based on z_{id} and z_{attr}. Previous researches [10] have shown that simple embedding concatenation may result in relatively fuzzy results. To solve the problem, Li [11] proposed novel *Adaptive Attentional Denormalization* (AAD) layers to improve feature integration in multiple levels. We employ cascaded n-AAD Residual Blocks in the generator to adjust attention regions of z_{id} and z_{attr} so that they can participate in synthesizing different parts.

6.3.2 Training Process

In training process, the identity encoder E_{id} is frozen while the others are trainable, where attributes encoder E_{attr} is trained to embed attribute representations disentangled from z_{id}, and the generator G is trained to reconstruct the original image with z_{id} and z_{attr}.

Given an input image X, the identity representations can be obtained as

$$z_{id} = E_{id}(X). \tag{6.10}$$

We use identity consistency loss $\mathcal{L}_{id}$ to make sure the identity of generated image $X' = G(z_{id}, z_{attr})$ still keeps the same.

$$\mathcal{L}_{id} = 1 - \frac{E_{id}(X') \cdot E_{id}(X)}{\|E_{id}(X')\|_2 \cdot \|E_{id}(X)\|_2}. \tag{6.11}$$

We also define attributes consistency loss $\mathcal{L}_{attr}$ which can be formulated as

$$\mathcal{L}_{attr} = \frac{1}{2} \sum_{k=1}^{n} \left\| z_{attr}^{k}(X') - z_{attr}^{k}(X) \right\|_2^2. \tag{6.12}$$

If the restored result X' is generated with the same z_{id} and z_{attr}, it should be as similar to the original image as we can. We set pixel-level $\mathcal{L}_2$ distance as reconstruction loss $\mathcal{L}_{rec}$,

$$\mathcal{L}_{rec} = \frac{1}{2} \|X' - X\|_2^2. \tag{6.13}$$

We take advantage of adversarial learning to train the framework and introduce adversarial loss $\mathcal{L}_{adv}$ to constrain the generated results indistinguishable from real images. To promote the image quality, it is necessary to expand the perception range of the discriminator, so we adopt m multiscale discriminators [12] for different resolution versions of the generated image X'_m.

$$\mathcal{L}_{adv}(X'_m, X_m) = \log\left(D(X_m)\right) + \log\left(1 - D(X'_m)\right). \tag{6.14}$$

Taking the above losses into account, the total loss function can be formulated as

$$\mathcal{L}_{total} = \mathcal{L}_{adv} + \lambda_1 \mathcal{L}_{id} + \lambda_2 \mathcal{L}_{attr} + \lambda_3 \mathcal{L}_{rec}, \tag{6.15}$$

where λ_1, λ_2, and λ_3 are the model tradeoff parameters.

6.3.3 Protection Process

In the protection phase, our approach takes the original image X, user-set password p, and privacy level parameter d as input. The goal is to generate a specific de-identification image with p and d whose identity has been protected while other attributes remain the same.

For the original image X, we firstly get identity embedding $z_{id} = E_{id}(X)$ and attributes embedding $z_{attr} = E_{attr}(X)$. The de-identification identity representation $z_{new} = M(z_{id}, p, d)$ can be formulated as

$$M(z_{id}, p, d) = \bar{z}_{id} \cdot \cos g(d) + z_{90} \cdot \sin g(d), \tag{6.16}$$

where

$$z_{90} = f(p) - (\bar{z}_{id} \cdot f(p)) \cdot \bar{z}_{id} \tag{6.17}$$

and $\bar{z}_{id}$ represents the normalized z_{id}.

Finally, we generate the de-identification result as

$$\mathcal{F}(X, p, d) = G(z_{new}, z_{attr}). \tag{6.18}$$

6.3.4 Recovery Process

In the recovery phase, our approach can restore the de-identified image $\mathcal{F}(X, p, d)$ to the original image X only when the right password is provided, which mainly differs from the protection process in the identity modification module M. Considering that when the password p is correct and d is not much different, the restored result of the same identity with the original image can still be obtained, so here we focus more on the correctness of passwords.

For the de-identified image $\mathcal{F}(X, p, d)$, we extract z_{new} and z_{attr} with the pretrained encoders. The restored identity embedding $\hat{z}_{id} = M^{-1}(z_{new}, p, d)$ can be calculated as

$$M^{-1}(z_{new}, p, d) = \frac{z_{new} - f(p) \cdot \sin g(d)}{\cos g(d) - A \cdot \sin g(d)}, \tag{6.19}$$

where

$$A = \frac{\cos^2 g(d) - (z_{new} - f(p) \cdot \sin g(d)) \cdot z_{new}}{\sin g(d) \cdot \cos g(d)} \tag{6.20}$$

and $z_{new} = E_{id}(\mathcal{F}(X, p, d))$. In fact, $A = \bar{z}_{id} \cdot z_r$, $\bar{z}_{id}$ is the normalized $z_{id} = E_{id}(X)$. The restored image $\hat{X}$ can be formulated as

$$\hat{X} = G(\hat{z}_{id}, z_{attr}). \tag{6.21}$$

6.4 Experiments

6.4.1 Implementation Details

Datasets We train the network using CelebA-HQ [13] dataset, which is derived from CelebA [14] containing 30k upscale images of celebrity faces. Randomly choose 27k images for training while the others for test. Each image has been aligned and cropped to 256×256 covering the whole face region. In addition, we also test the generalization ability on FFHQ [15] and CASIA-WebFace [16].

Experimental Settings We use the pretrained ArcFace [7] as identity encoder E_{id} and set the number of attribute representations $n = 8$ in Eq. (6.8). We train our network using Adam with $\beta_1 = 0$, $\beta_2 = 0.999$ and set the learning rate as 4×10^{-4}. The tradeoff parameters in Eq. (6.15) are set to $\lambda_{adv} = 0.1$, $\lambda_{id} = 5$, and $\lambda_{attr} = \lambda_{rec} = 10$. We define p as a six-digit password, each reference identity z_r is calculated by random $k = 10$ different identities, and define $f(p)$ as one-to-one mapping. Based on testing on CelebA-HQ and considering both privacy protection effectiveness and image quality and it cannot be restored when $\theta = 90°$, we define the relationship between θ and d as $g(d) = \begin{cases} 70 + d \times 5 & d \in [0, 4), \\ 70 + (d + 1) \times 5 & d \in [4, 9]. \end{cases}$

6.4.2 Evaluation Results

6.4.2.1 De-identification

Different Passwords We evaluate the diversity of our approach by generating different de-identification results with different passwords. The qualitative results are shown in Fig. 6.3. It can be seen that our method can transform the identity into different identities in a large range which is determined with the password p.

Different Privacy Level We evaluate the controllability by testing with different privacy levels d and present the qualitative results in Fig. 6.4. When d increases, the identity difference expands, while the de-identified results can still share a similar appearance with the original in general, and most of them have successfully deceived the face verification model. We will provide the quantitative evaluation in the following part.

Quantitative Evaluation We evaluate the performance of our approach from the perspectives of both privacy protection and image utility. Here we present the definition or explanation of the metrics we use.

(1) Privacy Protection: Almost all face verification models judge whether two images have the same identity by comparing identity embedding distance, so

Fig. 6.3 Various de-identification results. The leftmost column represents the original image, and the last five columns present diverse de-identified results with different passwords. Particularly, the images of each column share the same password

Fig. 6.4 The leftmost column represents the original image and the rest indicate the de-identified results with different privacy level (from left to right, the privacy level parameter d increases from 0 to 9)

that we define **identity distance (*Id-dis*)** and **successful protection rate (*SR*)** for protection effects evaluation. Id-dis indicates the distance between identity vectors e_{id} extracted from the face recognition model, which can be formulated as

$$Id - dis = D(e_{id}(X), e_{id}(\mathcal{F}(X, p, d)). \quad (6.22)$$

SR means the proportion of successful de-identification as

$$SR = 1 - \frac{1}{N}\sum_{i=1}^{N} f_{ver}(X, \mathcal{F}(X, p, d)), \tag{6.23}$$

when Id-dis $> \tau$, it considers two identities different as $f_{ver} = 0$ and otherwise $f_{ver} = 1$, and N is the number of testing. We respectively use the Face Recognition library, FaceNet trained on CASIA, and FaceNet trained on VGGFace2 for evaluation where the specific forms of D are all Euclidean distance.

(2) Image Utility: We define the rate at which faces in de-identified images can be detected as **face detectability (*DR*)**, as shown in Eq. (6.24), to measure the utility for computer vision tasks.

$$DR = \frac{1}{N}\sum_{i=1}^{N} f_{det}(\mathcal{F}(X, p, d)); \tag{6.24}$$

if the face can be detected, $f_{det} = 1$ and otherwise $f_{det} = 0$. We also detect face region and landmarks to calculate the **pixel-level distance (*pixel-dis*)** from the original image.

We randomly select several images from CelebA-HQ and de-identify them with random passwords p and privacy levels d. The privacy evaluation compared with DeepPrivacy [2] and Gu et al. [6] is represented in Table 6.1. It can be concluded that our method is more effective for identity protection with both larger identity distance and higher successful rate. We also generate the de-identification results using random passwords with each privacy level, and the variation of identity distance with d is shown in Fig. 6.5.

In Table 6.2, we apply computer vision algorithms on the de-identified images and compare the difference of pixel-level in face region, landmarks, eyes, nose, and mouth between the de-identification results and the original, as well as the detection rate of the de-identified. *Landmarks* indicates the mean distance of the total 68 keypoints, while *Eyes/Nose/Mouth* represents that of keypoints corresponding to each facial area. The utility evaluation proves that our method can guarantee the consistency of the face region and landmarks better, and most de-identified faces

Table 6.1 Privacy evaluation of de-identification results, where the values in the table indicate identity distance and successful de-identified rate *Id-dis/SR*. We choose the threshold of Face Recognition Library as $\tau = 0.6$ and the threshold of FaceNet as $\tau = 1.1$ according to [17]

	Face recognition	FaceNet (CASIA)	FaceNet (VGGFace2)
DeepPrivacy [2]	0.74623/0.939	1.19684/0.734	1.22889/0.816
Gu et al. [6]	**0.82234**/0.961	1.14419/0.704	1.16245/0.695
Ours	0.79195/**0.975**	**1.24421/0.913**	**1.27270/0.928**

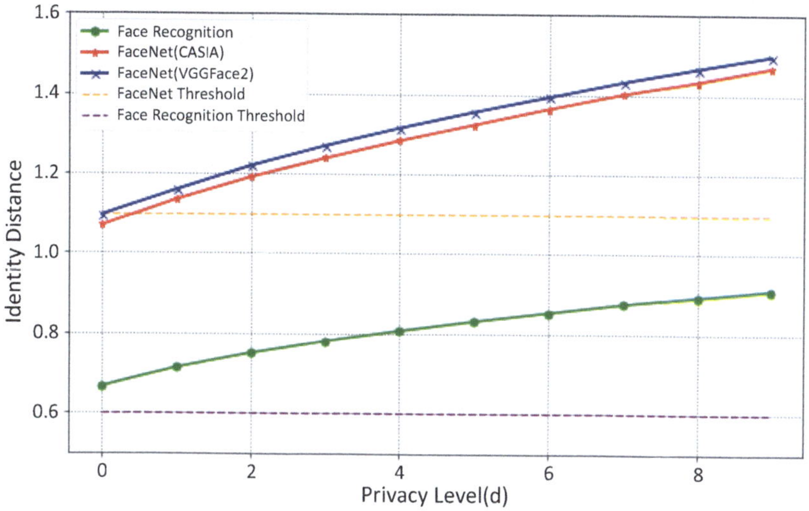

Fig. 6.5 Identity Distance(*Id-dis*). Larger distance illustrates better de-identification effects. When identity distance exceeds the threshold, the face verification model believes the identity has been varied

Table 6.2 Utility evaluation of de-identification results. The face region is detected with *OpenCV*, and landmarks are detected with *dlib*

		Pixel-dis↓				
	DR↑	Face	Landmarks	Eyes	Nose	Mouth
DeepPrivacy [2]	**1.0**	5.005	2.506	1.502	1.799	3.288
Gu et al. [6]	0.8585	0.925	2.346	1.810	1.906	2.139
Ours	0.9973	**0.225**	**1.969**	**1.236**	**1.546**	**1.900**

can be detected, which proves that it guarantees better utility for identity-agnostic computer vision tasks. We also show the tradeoff between privacy and utility in Fig. 6.6. Increasing the level of privacy protection will increase the pixel difference, which means the utility of the image will be reduced.

As shown in Fig. 6.7, compared with existing de-identification methods, our approach can retain more similarities with the original. Different from the generative adversarial network (GAN) conditioned on passwords proposed by Gu et al. [6], which needs to retrain the network for different passwords, our encryption process is relatively independent of the deep generative network, so that the password form can be defined more flexibly, the complexity will be reduced greatly, and the scope of identity changes can be infinitely expanded. Different from k-Same family algorithms [18–20] which can provide privacy guarantees and control privacy protection levels for the entire datasets, our method can control the extent of identity variation for each image.

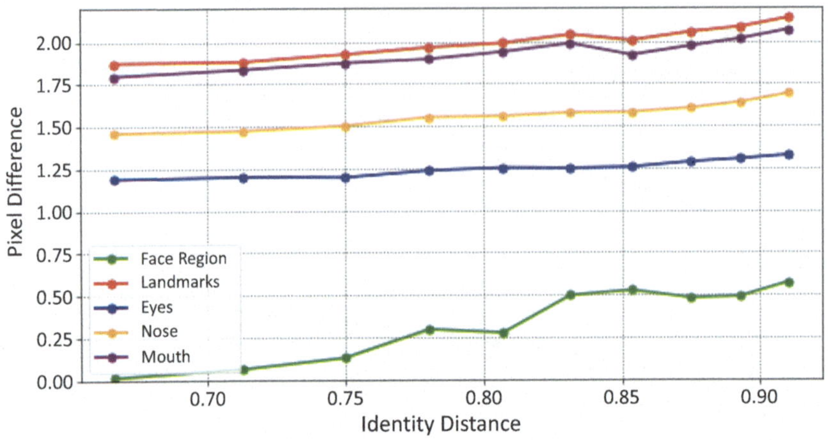

Fig. 6.6 The tradeoff between privacy and utility of the de-identified results. The abscissa represents the identity distance measured by the Face Recognition library, and the ordinate is the pixel difference of face region and keypoints

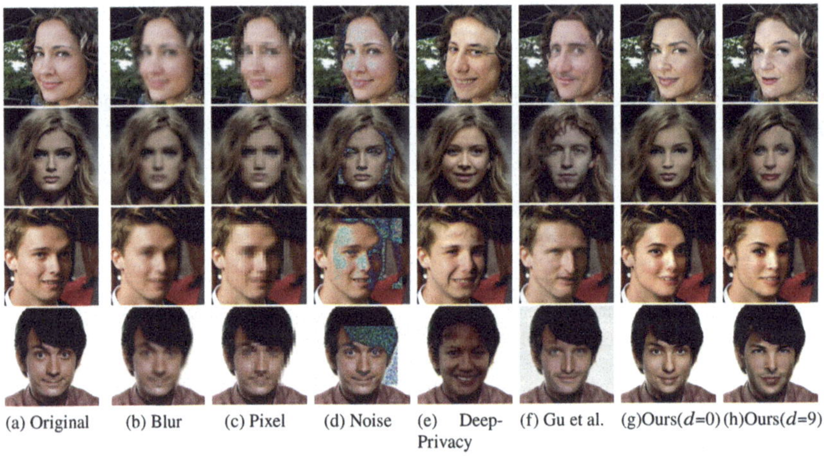

Fig. 6.7 De-identification results compared with existing methods, where ***(b), (c), and (d) are traditional methods and (e) and (f) are based on deep learning. From left to right: the original image, Gaussian Blur (s=8), pixelation (8×8), Gaussian noise (σ=15), DeepPrivacy [2], Gu et al. [6], and our de-identified results with the minimum and maximum privacy level d

6.4.2.2 Recovery

The restored results with correct or wrong passwords are presented in Fig. 6.8. When the attacker tries to recover the de-identified image with the wrong passwords, a good quality face image can still be obtained, but not the original identity information, which may confuse him/her and achieve more reliable protection.

Fig. 6.8 Recovery results. X: the original image, $p_{1,2}$: two different de-identified results, $p_m|p_n$: use p_n to restore the image de-identified with p_m, and $\hat{X}$: the correct recovery

While our framework is trained on CelebA-HQ, the generalization results tested on FFHQ and CASIA-WebFace are shown in Fig. 6.9, and it comes to the conclusion that our approach can apply to a wider range of images. In order to keep consistent with the model input, we first convert all test images to the size of 256×256 before feeding the model. The small artifacts are considered due to image distortion caused by interpolation or misalignment.

We compare de-identification results, wrong recovery, and correct recovery with [6] on both CelebA-HQ and CASIA shown in Figs. 6.10 and 6.11, which shows our de-identified results can retain more similarity with the original. Identity evaluation of incorrect and correct recovery are shown in Table 6.3, where the recovery is effective when using correct password, while wrong passwords will generate a different identity with a high probability. We evaluate the recovery quality in Table 6.4 using LPIPS (learned perceptual image patch similarity) [21] distance to measure perceptual similarity, PSNR (peak signal-to-noise ratio) and MAE (mean absolute error) to measure distortion at the pixel level, and SSIM (structural similarity) to measure the structure similarity. We compare our approach to three traditional methods and the method proposed by Gu et al. [6]. Specially, we deblur by Wiener filter, remove pixelation by bilinear interpolation, and denoise by nonlocal averaging. Based on the comparison, the restored images obtained by our method are the closest to the original with high image quality.

(a)X (b)p_1 (c)p_2 (d)p_3 (e)$\hat{Y}$ (f)$\hat{X}$

Fig. 6.9 FFHQ and CASIA generalization results with the model trained on CelebA-HQ. The upper two lines are from FFHQ, while the lower are from CASIA. X: original image, $p_{1,2,3}$: the de-identified results with three different passwords, $\hat{Y}$: wrong recovery, and $\hat{X}$: correct recovery

(a)p_1 (b)p_2 (c)$\hat{Y}$ (d)$\hat{X}$

Fig. 6.10 Compare results from CelebA-HQ with Gu et al. [6]. For the same input image, the upper row is our results, and the lower row is the results generated by Gu et al. [6]

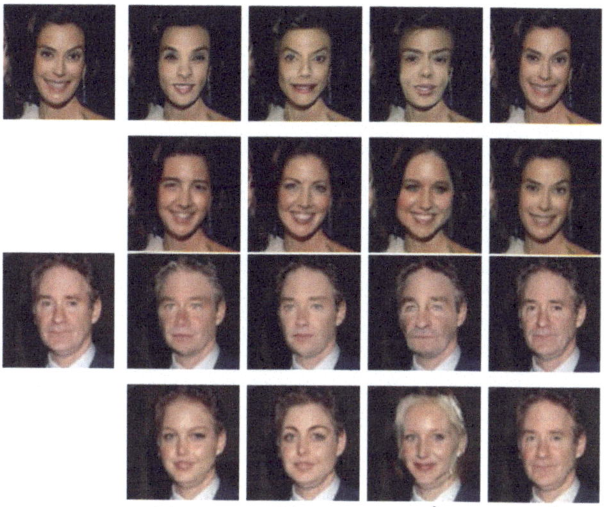

Fig. 6.11 Compare results from CASIA-WebFace with Gu et al. [6]. For the same input image, the upper row is our results, and the lower row is the results generated by Gu et al. [6]

Table 6.3 Id-dis/SR evaluation for incorrect/correct recovery

	Face recognition	FaceNet (CASIA)	FaceNet (VGGFace2)
Incorrect recovery	0.794/0.904	1.243/0.854	1.257/0.879
Correct recovery	0.228/0.035	0.368/0.035	0.401/0.035

Table 6.4 Comparison of the restored image quality

	LPIPS↓	PSNR↑	SSIM↑	MAE↓
Blur	0.242	**28.396**	0.802	**0.026**
Pixelation	0.447	23.159	0.671	0.040
Noise	0.264	22.163	0.701	0.046
Gu et al.	0.186	27.602	0.827	0.029
Ours	**0.062**	27.501	**0.902**	0.031

6.5 Conclusion

In this chapter, we propose a personalized and invertible de-identification method for privacy preservation. Our method first disentangles the representations of identity and attributes, encrypts or restores identity with latent space manipulation based on the password and the privacy level parameter, and finally reconstructs the de-identified or recovery image. In the protection phase, our approach can generate personalized de-identification results with different passwords and control the identity distance from the original by the privacy level parameter. In the recovery phase, our approach can restore if and only if the corresponding password is given, while the image with another identity will be generated when the attacker tries the wrong passwords. Experiments demonstrate the satisfactory performance in privacy protection and image utility of the de-identified results, as well as the quality of the restored, compared with the traditional or SOTA methods. Generalizing the proposed framework to handle face images of different resolutions and different poses is part of our future work. Besides, the de-identification in videos is also a problem worthy of research.

References

1. N. Vishwamitra, B. Knijnenburg, H. Hu, Y. P. Kelly Caine et al., Blur vs. block: Investigating the effectiveness of privacy-enhancing obfuscation for images, in *Proceedings of the IEEE Conference on Computer Vision and Pattern Recognition Workshops* (2017), pp. 39–47
2. H. Hukkelås, R. Mester, F. Lindseth, DeepPrivacy: a generative adversarial network for face anonymization, in *Advances in Visual Computing* (Springer, Berlin, 2019), pp. 565–578
3. M. Maximov, I. Elezi, L. Leal-Taixé, CIAGAN: conditional identity anonymization generative adversarial networks, in *Proceedings of the IEEE/CVF Conference on Computer Vision and Pattern Recognition* (2020), pp. 5447–5456
4. M. Gong, J. Liu, H. Li, Y. Xie, Z. Tang, Disentangled representation learning for multiple attributes preserving face deidentification. IEEE Trans. Neural Netw. Learn. Syst. **33**(1), 244–256 https://doi.org/10.1109/TNNLS.2020.3027617
5. Y. Nitzan, A. Bermano, Y. Li, D. Cohen-Or, Face identity disentanglement via latent space mapping. ACM Trans. Graph. **39**, 1–14 (2020)
6. X. Gu, W. Luo, M. S. Ryoo, Y. J. Lee, Password-conditioned anonymization and deanonymization with face identity transformers, in *European Conference on Computer Vision* (2020)
7. J. Deng, J. Guo, N. Xue, S. Zafeiriou, ArcFace: additive angular margin loss for deep face recognition, in *Proceedings of the IEEE/CVF Conference on Computer Vision and Pattern Recognition (CVPR)* (2019)
8. H. Wang, Y. Wang, Z. Zhou, X. Ji, D. Gong, J. Zhou, Z. Li, W. Liu, CosFace: large margin cosine loss for deep face recognition, in *2018 IEEE/CVF Conference on Computer Vision and Pattern Recognition* (2018), pp. 5265–5274
9. W. Liu, Y. Wen, Z. Yu, M. Li, B. Raj, L. Song, SphereFace: deep hypersphere embedding for face recognition, in *2017 IEEE Conference on Computer Vision and Pattern Recognition (CVPR)* (2017), pp. 6738–6746
10. J. Bao, D. Chen, F. Wen, H. Li, G. Hua, Towards open-set identity preserving face synthesis, in *2018 IEEE/CVF Conference on Computer Vision and Pattern Recognition* (2018), pp. 6713–6722
11. L. Li, J. Bao, H. Yang, D. Chen, F. Wen, FaceShifter: towards high fidelity and occlusion aware face swapping (2019). arXiv preprint arXiv:1912.13457
12. W. Tang, G. Li, X. Bao, T. Li, MSCGAN: multi-scale conditional generative adversarial networks for person image generation, in *2020 Chinese Control And Decision Conference (CCDC)* (2020), pp. 1440–1445
13. T. Karras, T. Aila, S. Laine, J. Lehtinen, Progressive growing of GANs for improved quality, stability, and variation, in *International Conference on Learning Representations* (2018). https://openreview.net/forum?id=Hk99zCeAb
14. Z. Liu, P. Luo, X. Wang, X. Tang, Deep learning face attributes in the wild, in *Proceedings of International Conference on Computer Vision (ICCV)* (2015)
15. T. Karras, S. Laine, T. Aila, A style-based generator architecture for generative adversarial networks, in *Proceedings of the IEEE/CVF Conference on Computer Vision and Pattern Recognition (CVPR)* (2019)

16. D. Yi, Z. Lei, S. Liao, S. Z. Li, Learning face representation from scratch (2014). ArXiv, abs/1411.7923
17. F. Schroff, D. Kalenichenko, J. Philbin, FaceNet: a unified embedding for face recognition and clustering, in *Proceedings of the IEEE Conference on Computer Vision and Pattern Recognition (CVPR)* (2015)
18. R. Gross, E. Airoldi, B. Malin, L. Sweeney, Integrating utility into face de-identification, in *PET'05 Proceedings of the 5th international conference on Privacy Enhancing Technologies* (2005), pp. 227–242
19. R. Gross, L. Sweeney, F. De la Torre, and S. Baker, Model-based face de-identification, in *2006 Conference on Computer Vision and Pattern Recognition Workshop (CVPRW'06)* (2006), pp. 161–161
20. E. Newton, L. Sweeney, B. Malin, Preserving privacy by de-identifying face images,. IEEE Trans. Knowl. Data Eng. **17**(2), 232–243 (2005)
21. R. Zhang, P. Isola, A.A. Efros, E. Shechtman, O. Wang, The unreasonable effectiveness of deep features as a perceptual metric, in *Proceedings of the IEEE Conference on Computer Vision and Pattern Recognition (CVPR)* (2018)

Chapter 7
High Quality Face De-identification with Model Explainability

7.1 Introduction

Concerns about individual private information disclosure are growing with the development of computer vision techniques and image understanding applications. Face de-identification is a process that aims to remove all identification information of the person from an image, while maintaining as much information on the action and its context [1]. Ideally, while the identity information is protected, other identity-agnostic features (e.g., pose, expression, and background) will not be affected. The de-identified images can still be used for identity-agnostic tasks, such as face detection and expression recognition. Accordingly, great efforts are paid to achieve an effective privacy–utility tradeoff [2–9]. Face de-identification can allow individuals to share personal portraits with confidence, while eliminating some ethical and legal restraints on facial data releasing.

Early face de-identification methods carry out various obfuscation operations on detected private area, which seriously impair the image's ornamental value and are not reliable when facing advanced face recognition tools [10]. K-same family methods [11–13] are once hot, but they are restrained by their strict using conditions. At present, there are two main types of methods. One kind uses adversarial noise to generate de-identified faces that can be visually indistinguishable from the original one [14–16]. However, they are highly dependent on the accessibility to target systems and lack generalization ability. The other kind exploits GANs to disentangle, manipulate, and finally protect identity features in the latent spaces [2–9]. These methods make great efforts to achieve the balance between privacy and utility through a network in the manner of an adversarial training. The results

Main contents of this chapter have been published in "Wen, Y., Liu, B., Cao, J., Xie, R., & Song, L. (2023). Divide and Conquer: a Two-Step Method for High Quality Face De-identification with Model Explainability. In Proceedings of the IEEE/CVF International Conference on Computer Vision (pp. 5148–5157)."

Y. Wen et al., *Face De-identification: Safeguarding Identities in the Digital Era*,
https://doi.org/10.1007/978-3-031-58222-6_7

depend heavily on the degree of latent space disentanglement, which is neither clear nor satisfactory. Besides, most existing methods are designed in constrained scenarios and do not work well with various poses and expressions, which also need to be improved.

Unlike previous works, we aim to break away from this traditional privacy–utility tradeoff in face de-identification studies and instead provide a reliable and explainable method of protecting individual identities. Our inspiration for this approach stems from the observation that wearing a human skin mask can effectively change one's identity. This realization highlights that a convincing de-identification requires substantial changes to the overall geometry of the facial features such as eyes, nose, ears, mouth, and facial bones. Such transformations are practically impossible to achieve with mere makeup or even surgical procedures (since that surpass the physical limits of the human body). In contrast, hairstyle, accessories, and skin color are examples of identity-agnostic features that can be easily altered by a stylist. However, they significantly impact the human perception of visual similarity between two faces. Thus, we contend that protecting privacy and retaining utility can be two distinct objectives that necessitate different strategies. By separating these objectives, we can focus on each objective independently to achieve better results (Fig. 7.1).

Our proposed solution, IDeudemon, adopts the "divide and conquer" strategy to achieve privacy protection and utility preservation in two distinct steps. In the first step, we use a 3D parametric modelling approach to estimate the facial geometry and obfuscate the face's 3D identity representation to conceal the real identity. Specifically, we begin by leveraging a monocular face reconstruction network to approximate the coarse 3D parameters of the given face. Using this initialization, we employ an NeRF model to calculate the face's accurate 3D parameters (ID code, appearance code, and camera code). Subsequently, we apply a protective perturbation to the real ID code to get the protected ID code. Finally, the NeRF model renders an identity-protected fitted face, which has a significant change in the facial features' geometric structure.

In the second step, we focus on producing high quality images based on the fitted face, which is neither natural nor realistic. We first use face parsing maps to preserve the identity-agnostic features and maintain the visual similarity with the original image as much as possible. Then, we train a GAN to restore the de-identified face with realistic details by referring to generative facial priors. Finally, we can acquire high quality visual-pleasing de-identified results.

Our main contributions are described as follows:

- We propose IDeudemon, a novel two-step NeRF-based method for face de-identification. Instead of achieving privacy–utility tradeoff in one network adversarially, for the first time, we divide privacy protection and utility preservation into two separate steps. IDeudemon can protect identity without weighing the image utility at the same time and has good explainability [17].
- We confuse the real identity by a 3D parametric NeRF model, which modifies the facial geometry and changes the identity. Hence, our method has excellent

Fig. 7.1 IDeudemon for face de-identification at different resolutions. (a) 256 × 256, (b) 512 × 512, (c) 1024 × 1024. In each pair, left is the original image and right is the corresponding de-identified result. The results show that face identities are changed in a perceptually natural manner, while all other characteristics (hairstyles, accessories, backgrounds, poses, expressions, etc.) remain the same

privacy performance and this process is explainable. The definition of the identity refers to the mature 3D prior from 3DMMs and is refined by the NeRF model. This verified disentangled identity code makes IDeudemon well preserve nonidentity features, such as expression, pose, and illumination.

- We propose a second step to intently restore high quality faces based on the fitted results of NeRF. We devise visual similarity assistance to retain identity-agnostic features and train a GAN to generate realistic facial details. These designs lead to good utility performance.

- Experimental results on two diverse face datasets (ethnicity, age, etc.) have shown the effectiveness of our proposed IDeudemon. In particular, our method brilliantly maintains the original poses and expressions and can achieve face de-identification on megapixels.

7.2 Related Work

7.2.1 3D Monocular Face Reconstruction

3D monocular face reconstruction refers to reconstructing the 3D model of a face from a 2D image. Methods [18–20] based on 3D Morphable Models (3DMMs) [21] have dominated this field. Besides, there exist some methods advocating direct model-free reconstruction [22] or based on other innovative models [23]. However, all these methods suffer from the problem that the reconstructed faces are not realistic. Recently, NeRF shows encouraging results in capturing implicitly encoded complex scene structures and fitting 3D-consistent images with fine details [24–26]. As faces contain regular 3D structure, NeRF-based 3D face modelling researches [27–30] are now in full swing.

7.2.2 Blind Face Restoration

Face images are often degraded due to complicated factors in real word, and blind face restoration (BFR) aims at recovering high quality faces from the low-quality counterparts suffering from unknown degradation [31]. Even for today's powerful GANs, imagining reasonable details out of thin air is too difficult. Current BFR methods always require facial priors, which can be coarsely categorized into three types according to the sources: geometric priors [32, 33], reference priors [34–37], and generative priors [31, 38–41]. Among them, the third kind is not limited by the quality of corrupted faces, the accessibility of high-resolution references having the same identity, or the capacity of the references. So it is the most suitable for the restoration of fitted faces rendered by current NeRF.

7.3 Methodology

7.3.1 Overview of IDeudemon

Given an input face image X without any protection, the purpose of face de-identification is to generate a photo-realistic image X' that conceals the real identity.

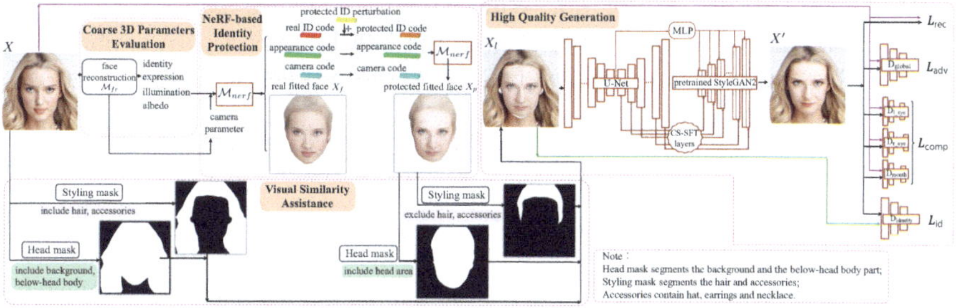

Fig. 7.2 The architecture of IDeudemon. To protect identity, we first estimate the coarse 3D parameters of input image X as an initialization. Then an NeRF model is employed to calculate X's accurate 3D codes and fitted face X_f. After adding protective perturbation to the real ID code, the NeRF generates the de-identified fitted face X_p. To preserve utility, we design visual similarity assistance to directly retain the identity-agnostic areas and train a GAN referring to generative priors to produce the final high quality de-identified face X'

The de-identified face X' is visually similar to the original image X but should be judged as a different person by recognition tools when comparing with X.

Figure 7.2 illustrates the overall pipeline of the proposed IDeudemon, which protects privacy and guarantees utility in distinct steps sequentially. In the following, we discuss the two steps in detail.

7.3.2 Step I: Parametric Identity Protection

Coarse 3D Parameters Evaluation 3DMMs are generative parametric models for the 3D representation of human faces. They are built from a set of 3D facial scans, coupled to each other with anatomical correspondences, and can represent any unseen faces as a linear combination of the training set [42]. Fitting 3DMMs, also known as 3D face reconstruction, facilitates the estimation of identity, pose, albedo, and illumination-related parameters from the face images. In order to provide a good basis for real-time NeRF-based fitting, we employ a 3DMM model [43], denoted as $\mathcal{M}_{fr}$, to initialize the 3D parameters [44] of the input face image X, which is denoted as

$$c_{id}, c_{exp}, c_{alb}, c_{illu} = \mathcal{M}_{fr}(X). \tag{7.1}$$

c_* represent the coarse 3DMM parameters for four disentangled factors: identity c_{id}, expression c_{exp}, and albedo c_{alb} of the face X, and the illumination c_{illu} of the scene. These parameters are initialized by solving an inverse rendering optimization [45] based on the 3DMM model [43]. Although the initial identity parameter only describes the coarse geometry of the face area (without hair, teeth, etc.), it will be adaptively adjusted and become accurate through the NeRF model described below.

NeRF-Based Identity Protection With initialized 3DMM parameters c_*, we employ a pretrained parametric NeRF model [30], denoted as $\mathcal{M}_{nerf}$, to obtain the accurate 3D parameters and the fitted face X_f of original image X:

$$X_f, z_{id}, z_{app}, z_{cam} = \mathcal{M}_{nerf}(X, c_{id}, c_{exp}, c_{alb}, c_{illu}, C). \tag{7.2}$$

X_f is the fitted image. C is the camera parameter used for rendering (detailed calculation is shown in [30]). z_* represent the computed 3D codes for face image X, whose dimensionality is the same as that of the corresponding coarse 3D parameters. In particular, because our de-identification task hopes to distinguish the identity feature from all other facial features, we let z_{id} represent the identity separately and name it as ID code. Then we let the appearance code z_{app} contain not only the expression and albedo of the face in x, but also the illumination of the whole scene. In addition, as the density field from NeRF can implicitly encode the 3D geometry of the scene, we can also acquire a camera code z_{cam}, which reflects the pose of the face in X.

To protect the real identity information, we use a noise generator to generate benign Gaussian noise n whose size equals to the fitted ID code z_{id} according to the actual requirements. Then we directly add the protective noise on z_{id} to get a perturbed ID code z'_{id}:

$$z'_{id} = z_{id} + n. \tag{7.3}$$

In Sect. 4.2, we perform a series of perturbation analysis experiments, where we get the optimum scale range of perturbation for identity protection.

At the end of this step, the NeRF model takes the protected identity code z'_{id}, the original appearance code z_{app}, and camera code z_{cam} as input and fits the final identity-protected fitted face X_p. It is formulated as

$$X_p, z'_{id}, z_{app}, z_{cam} = \mathcal{M}_{nerf}(z'_{id}, z_{app}, z_{cam}). \tag{7.4}$$

Since our parametric NeRF model refers to the 3DMM model, the whole de-identification process has good explainability. Moreover, since the perturbation is directly added on the disentangled ID code, the result with faithful identity change still well retains identity-agnostic features (i.e., expression, albedo, illumination, and pose).

7.3.3 Step II: Utility Preservation

Despite the promising de-identified fitted result X_p of parametric NeRF model, it has limitations in terms of realistic looks. In order to generate visual-pleasing high quality faces, we take several measures as follows.

Visual Similarity Assistance As mentioned earlier, hairstyles, accessories, and background are weakly related to the identity but may occupy a pretty large space and greatly affect human perception of visual similarity and the subsequent use. Therefore, we use face parsing maps [46, 47] to generate a head mask (which segments the background and the below-head body part) and a styling mask (which segments the hair and accessories) for X and X_p. Here we combine the hair, accessories, background, and below-head body section in the original image X with the segmented face except for the hair and accessories in the fitted image X_p. Therefore, a hybrid face image X_l is produced, which conceals the real identity and retains the identity-agnostic areas. As seen in Fig. 7.2, X_l has realistic identity-agnostic features, low-quality face regions, and some irregular white gaps, which still need to be improved.

High Quality Generation The translation from hybrid image X_l to desired high quality de-identified photo X' aims to accomplish a face restoration task, which transforms degraded image to its photo-realistic counterpart with distinct and discernible details. The domain gap is pretty large, so this task is challenging. Thanks to the leaps and bounds in BFR, here we employ a publicly available GAN model [31] that leverages rich and diverse priors encapsulated in the pretrained StyleGAN2 [48] to achieve high quality de-identified face generation. This GAN model is mainly composed of two parts: a U-Net [49] which is responsible for removing degradation and extracting "clean" features of X_l, and a pretrained StyleGAN2 that provides facial priors. They are bridged by a latent code mapping and several Channel-Split Spatial Feature Transform (CS-SFT) layers in a coarse-to-fine manner. By training this GAN model, we can obtain high quality de-identified image X'.

IDeudemon enjoys the benefits of separating the implementation of protecting privacy and preserving utility, so has the advantage of adjusting the degree of identity protection as practical need while maintaining remarkable utility performance. Our approach no longer needs to struggle with the annoying tradeoff between privacy and utility.

7.3.4 Loss Function

We train the GAN model with triplet of images X, X_l, and X'. We inherit the validated loss functions from [31] and adjust them as the requirements of our mission.

Reconstruction Loss The widely used $\mathcal{L}_1$ loss and perceptual loss are summed as the reconstruction loss $\mathcal{L}_{rec}$ [50, 51], which targets at making the output X' look like the original face X:

$$L_{rec} = \lambda_{l_1} \left\| X' - X \right\|_1 + \lambda_{per} \left\| \phi(X') - \phi(X) \right\|_1, \tag{7.5}$$

where ϕ is the pretrained VGG-19 network [52], and we select the $conv1, \cdots, conv5$ feature maps before activation.

Adversarial Loss The adversarial loss $\mathcal{L}_{adv}$ is responsible for restoring realistic textures, enforcing generated faces to be indistinguishable from real images. It is formulated as

$$L_{adv} = -\lambda_{adv}\mathbb{E}_{X'}[softplus(D(X'))], \tag{7.6}$$

where D denotes the discriminator and λ_{adv} represents the adversarial loss weight.

Facial Component Loss Given that people easily detect mistakes in the appearance of a human face (uncanny valley effect), we also use the facial component loss with local discriminators for left eye, right eyes, and mouth, which is defined as follows. The first term is the discriminative loss [53], and the second term is the feature style loss [54]:

$$\begin{aligned} L_{comp} = \sum_{ROI} \lambda_{local}\mathbb{E}_{X'_{ROI}}[log(1 - D_{ROI}(X'_{ROI}))] + \\ \lambda_{f_s} \left\| Gram(\psi(X'_{ROI})) - Gram(\psi(X_{ROI}))\right\|_1, \end{aligned} \tag{7.7}$$

where ROI is region of interest [55] from the component collection, which includes the $left\_eye$, the $right\_eye$, and the $mouth$. D_{ROI} is the local discriminator for each region. The feature style loss attempts to match the Gram matrix statistics [56] of real and restored patches from multiple layers of the learned local discriminators, which has been demonstrated to be conductive to generating realistic facial details and reducing unpleasant artifacts. Besides, ψ denotes the multiresolution features from the learned discriminators. λ_{local} and λ_{f_s} represent the loss weights of local discriminative loss and feature style loss, respectively.

Identity Preserving Loss During the process of high quality generation, the "fake" identity generated in the previous step, i.e. the identity of X_l, must remain as constant as possible. We employ a pretrained state-of-the-art (SOTA) face recognition model [58] to extract identity features. Deng et al. [58] is chosen because it can provide highly discriminative identity features and has a clear geometric interpretation due to the exact correspondence to the geodesic distance on the hypersphere. We use the identity preserving loss $\mathcal{L}_{id}$ to ensure that the identity of X' is the same as X_l:

$$\mathcal{L}_{id} = \lambda_{id}\left(1 - \frac{r_{id}(X') \cdot r_{id}(X_l)}{\|r_{id}(X')\|_2 \cdot \|r_{id}(X_l)\|_2}\right), \tag{7.8}$$

where r_{id} represents the identity feature extract by Deng et al. [58]. λ_{id} denotes the weight of identity preserving loss. Here we use cosine similarity rather than the original $\mathcal{L}_1$ distance in [31] because we think it better fits the angular margin based identity extractor [58] (and is proved in Sect. 4.4).

The overall model objective is a combination of the above losses:

$$L_{total} = L_{rec} + L_{adv} + L_{comp} + L_{id}. \tag{7.9}$$

The hyperparameters are set as follows: $\lambda_{l_1} = 0.1$, $\lambda_{per} = 2$, $\lambda_{adv} = 0.1$, $\lambda_{f_s} = 200$, and $\lambda_{id} = 5$.

7.4 Experiments

7.4.1 Experimental Setup

Datasets We choose the FFHQ dataset [59], which contains 70K high-resolution face images with diverse demographic information like age, gender, and race, to train our GAN model in Step II. We randomly select 60K images for training and 10K for testing. All images are aligned and cropped to size 512 × 512 covering the whole face, as well as some background regions. Moreover, in order to compare with other methods fairly, we also test IDeudemon on the CelebA–HQ dataset [60] and show our generalization ability (see Sect. 4.3 for details).

Evaluation Metrics We evaluate the proposed IDeudemon in terms of two metrics, as described below:

(1) Privacy metrics. Following previous work [6], we measure the $\mathcal{L}_2$ distance of embedding vectors from the de-identified and original faces extracted by a pretrained face recognition model, denoted as **DIS**, to evaluate the quality of identity protection. For a fair comparison, we employ two models that are excluded from our training, i.e., the Face Recognition library[1] (denoted as FR), and the FaceNet [57] that is pretrained on two public datasets (CASIA-Webface [61] and VGGFace2 [62]), respectively.
(2) Utility metrics. We evaluate not only the quality of the de-identified images, but also the retention ability to pose and expression. Specifically, **PSNR**, **SSIM**, and **FID** are chosen to evaluate the generation quality. PSNR and SSIM are widely used objective methods to measure the difference between two images, while FID can measure the distance between the generated distribution and the real distribution. Besides, the $\mathcal{L}_2$ distances between pose and expression vectors from the de-identified and original faces extracted by an open-sourced pose estimator [63] and a 3D facial model [64] are calculated as pose (denoted as **POSE**) and expression (denoted as **EXP**) similarity.

Implementation Details We implement our framework as shown in Fig. 7.2. Since the value range of the ID code is between $[-1, 1]$, after Step I, the part out of the

[1] https://github.com/ageitgey/face_recognition

range needs to be truncated to ± 1, depending on which value is closer. The sizes of different facial codes are $c_{id}, z_{id} \in \mathbb{R}_{100}$, $c_{exp} \in \mathbb{R}_{79}$, $c_{alb} \in \mathbb{R}_{100}$, $c_{illu} \in \mathbb{R}_{27}$, and $z_{app} \in \mathbb{R}_{206}$, respectively. During the training of the GAN model in Step II, the minibatch size is set to 6. We augment the training data with horizontal flip and color jittering. We train our model with Adam optimizer [65] for a total of 300k iterations. The learning rate was set to 2×10^{-3} and then decayed by a factor of 2 at the 220k-th, 270k-th iterations.

7.4.2 Protective Perturbation Analysis.

This section analyzes the performance of our IDeudemon with different levels of perturbation applied on the original ID code in Step I. The additive Gaussian noise n is sampled from a normal distribution. The *loc* is set to 0, the value of its *scale* belongs to $\{0.05, 0.10, 0.15, 0.20, 0.25, 0.30, 0.35, 0.40\}$, and the size equals to z_{id}. Ten de-identified faces are generated for every test face image under each *scale* value. Various statistical mean metric results are calculated at each *scale* value.

Figure 7.3 shows the qualitative results. It can be observed that with the increase of the noise *scale*, the geometric difference between the de-identified and original

Fig. 7.3 Qualitative results of the influence of the noise *scale* on the **FFHQ**. The first column shows the original face images. The rest columns demonstrate de-identified faces whose identity distances are closest to the mean distance under every *scale* value

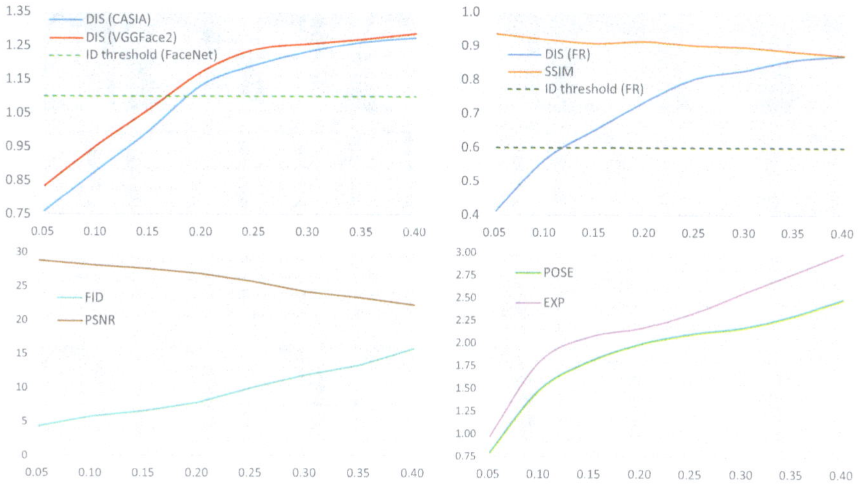

Fig. 7.4 The de-identified performance variation with respect to the noise *scale* on the **FFHQ**. The x-axis indicates the *scale* value, and the y-axis indicates different metric values. The identity judgment threshold is 0.6 for Face Recognition library [6] and 1.1 for FaceNet [57]

faces expands, while the identity-agnostic attributes (hairstyle, background, etc.) are still maintained. The quality of the de-identified images is consistently good and is almost comparable to the quality of the original images. All synthetic images have sharp details such as eyelashes, wrinkles, teeth, and lips. Quantitative results are shown in Fig. 7.4. One can see that the degree of identity protection can be adjusted, along with the change of utility. It is worth noting that the utility is kept at a good level (e.g., the FID values are always low). Particularly, we note that when the noise *scale* is smaller than 0.2, the results are too similar to the original faces and the ability to protect identity is not strong; when the *scale* is larger than 0.3, the geometric structure of the faces begins to become exaggerated (such as eccentric eyes, noses, wrinkles, and shadows).

Based on the extensive experiments mentioned above, taking into account the visual effects and evaluation metrics comprehensively, we recommend the users to set the *scale* of the protective perturbation between 0.2 and 0.3 to obtain de-identified faces efficiently with well-preserved appearance. We no longer show the case of adding Gaussian noise with larger *scale* values because the generated faces will be quite visually exaggerated.

7.4.3 Comparison with SOTA Methods.

To validate the effectiveness of the proposed IDeudemon, we compare it with several SOTA de-identification methods: DeepPrivacy [2], AnonymousNet [4], CIAGAN

Fig. 7.5 Qualitative comparison on the **CelebA-HQ** for face de-identification. Our IDeudemon conceals the real identity and produces photo-realistic details at the same time. **Zoom-in for best view**

[3], Gu et al. [5], Cao et al. [6], and AMT-GAN [16]. For fairness, the test dataset is CelebA–HQ [60], and all images are aligned and cropped to size 256×256.

To test on the dataset, we first bilinearly interpolate the input image to 512×512 and then process it according to the pipeline in Fig. 7.2. Because (1) the NeRF-based 3D fitting in Step I can still handle the image without photo-realistic details, (2) the GAN model in Step II is trained to process this kind of degradation, our de-identification results are still outstanding in terms of generation quality. The *scale* of protective Gaussian noise is set to 0.25. The final outputs are rescaled to 256×256, covering the whole face, as well as some background regions.

Qualitative results are shown in Fig. 7.5a. One can see that the competing methods fail to produce photo-realistic faces, especially when the original face has a large pose (the last two rows) or expression (the second row). In contrast, our IDeudemon obfuscates the human identities in a perceptually natural manner; meanwhile, the de-identified face still shares similar appearance, as well as the same pose, expression, illumination, and background with the original face. It is worth noticing that our results are high fidelity and can retain clear lips, teeth, and even eyelashes, which is superior to other methods.

Quantitative results are presented in Table 7.1. Our method obtains the best scores in privacy metrics, clearly confirming our initial motivation that manipulating the 3D parametric ID code can greatly benefit the identity protection. One can see that our IDeudemon achieves comparable PSNR and SSIM indexes to other competing methods but achieves significantly better results on FID index, which is a better measure for the image perceptual quality. In addition, our method outperforms the other methods in retaining pose and expression. These verify the efficiency of

Table 7.1 Quantitative comparison with SOTA methods on the **CelebA-HQ**. ↑ means higher is better, and ↓ means lower is better. Red and blue indicate the best and the second best performance

Method	DIS↑			PSNR↑	SSIM↑	FID↓	POSE↓	EXP↓
	FR	CASIA	VGGFace2					
DeepPrivacy [2]	0.783	1.091	1.187	21.3	0.791	24.6	6.22	5.27
AnonymousNet [4]	0.497	0.875	0.936	20.4	0.803	53.7	3.69	4.02
CIAGAN [3]	0.671	0.919	1.085	18.6	0.522	28.1	8.93	5.19
Gu et al. [5]	0.812	1.207	1.224	23.1	0.751	39.7	3.95	3.96
Cao et al. [6]	0.794	1.206	1.231	24.1	**0.902**	22.6	3.04	2.81
AMT-GAN [16]	0.596	0.927	0.941	21.0	0.799	33.3	3.02	2.86
IDeudemon	**0.819**	**1.228**	**1.233**	**25.9**	0.898	**8.7**	**2.96**	**2.79**

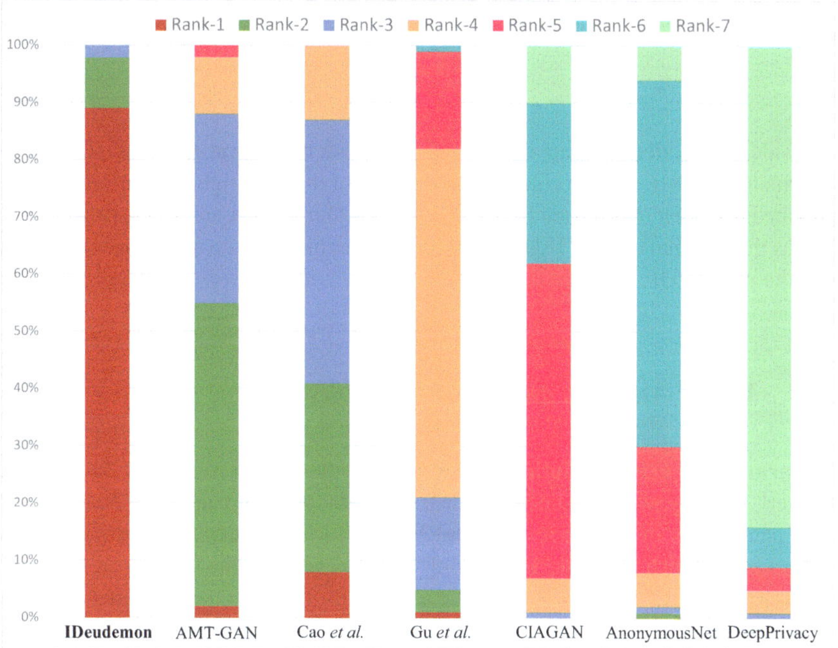

Fig. 7.6 User study results of different de-identification methods

our designs in ensuring utility and make IDeudemon have the least impact on the subsequent use of the de-identified images.

User Study The de-identified results of comparison methods and our IDeudemon on 100 face images are presented in a random order to 10 volunteers for subjective evaluation. The volunteers are asked to rank the 7 de-identified outputs of each input image according to their perceptual quality. Finally, we collect 7k votes, and the statistics are presented in Fig. 7.6b. As can be seen, our IDeudemon receives much more rank-1 votes than other SOTA methods.

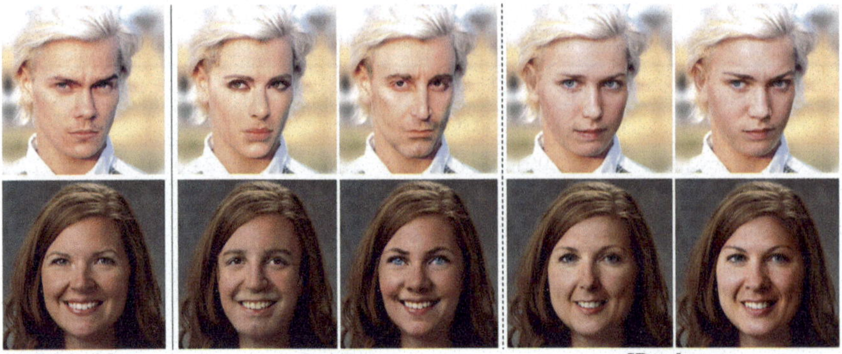

Fig. 7.7 Comparison with StyleFace [8] at megapixel level (1024 × 1024, from the paper sample image)

Besides, IDeudemon can conduct face de-identification at megapixel level (inherits from [48]), and we compare it with one of the first high-resolution methods published last year, StyleFace [8] (see Fig. 7.7). Our results are at least visually as good as the original ones of [8], despite having to run on the cropped faces extracted from the paper PDF.

7.4.4 Model Analysis and Ablation Study

3D Parametric Fitting Method Selection In the first step of our "divide and conquer" strategy, what we need is a fast, accurate tool that can fit the disentangled facial parameters in 3D space. The NeRF model [30] created last year is the first work to accomplish this task. Hong et al. [30] has verified the validity of each part and its SOTA fitted effect. Therefore, we adopt it for face parametric fitting in Step I. The brilliant de-identification effects of IDeudemon have proven the correctness of this choice.

Ablation Study of Step II In order to validate the effectiveness of our various designs in Step II, in this section we conduct an ablation study by introducing some variants of our IDeudemon and comparing their performance.

We first pick and train five SOTA face restoration models to respectively replace the GAN model [31] we used as five variants. They are denoted as BOPB [40], GPEN [41], RestoreFormer [35], CodeFormer [37], and VQFR [36]. Then w/o vsa refers to the IDeudemon model without visual similarity assistance. Additionally, we validate the necessity of the loss functions, which are indicated as w/o L_{rec}, w/o L_{adv}, w/o L_{comp}, and w/o L_{id}. We specifically calculate the identity preserving loss by using $\mathcal{L}_1$ distance (like [31]) rather than cosine similarity and denote it as *idloss*.

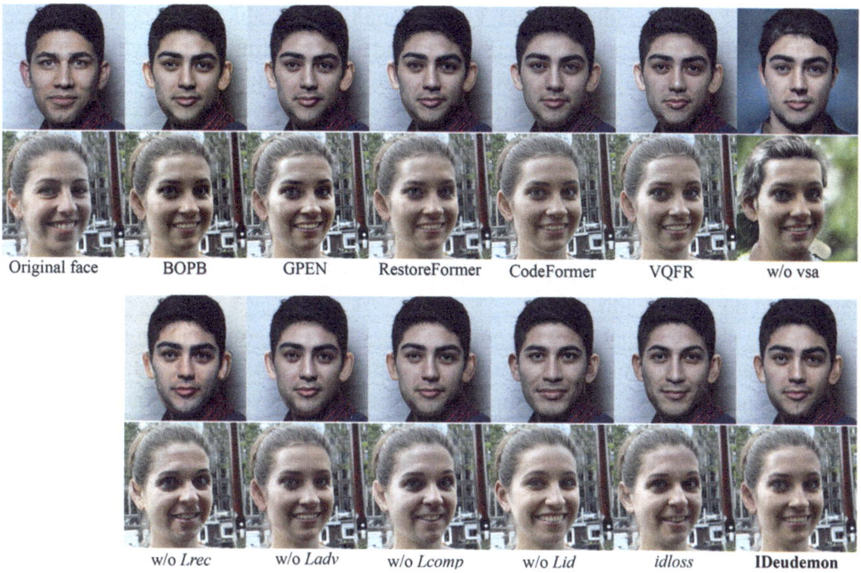

Fig. 7.8 Ablation studies on GAN model, visual similarity assistance, and identity preserving loss on the **FFHQ**. **Zoom-in for best view**

Table 7.2 Ablation study results of Step II on the **FFHQ**. ↑ means higher is better, and ↓ means lower is better. Red and blue indicate the best and the second best performance

Method	DIS↑			PSNR↑	SSIM↑	FID↓	POSE↓	EXP↓
	FR	CASIA	VGGFace2					
BOPB [40]	0.803	1.083	1.224	26.2	0.899	17.63	2.963	2.783
GPEN [41]	0.794	1.186	1.236	24.8	0.895	11.65	2.975	2.772
RestoreFormer [35]	0.802	1.191	1.239	24.6	0.889	11.14	2.956	2.768
CodeFormer [37]	0.801	1.189	1.236	23.8	0.905	10.83	2.950	2.778
VQFR [36]	0.796	1.185	1.238	24.2	0.898	11.95	3.006	2.839
w/o vsa	**0.815**	**1.193**	**1.244**	20.4	0.728	25.78	3.084	3.854
w/o L_{rec}	0.801	1.188	1.236	24.2	0.847	10.07	3.112	2.847
w/o L_{adv}	0.799	1.191	1.231	24.6	0.863	11.53	2.993	2.788
w/o L_{comp}	0.803	1.189	1.237	25.4	0.891	10.12	2.947	2.831
w/o L_{id}	0.417	0.816	0.965	**26.3**	**0.912**	**9.613**	2.973	**2.754**
idloss	0.768	1.079	1.203	25.5	0.901	10.06	2.958	2.762
IDeudemon	0.804	1.192	1.239	25.8	0.903	9.99	**2.942**	2.761

We perform on the FFHQ dataset to evaluate IDeudemon and its seven variants. After the common Step I, except that w/o vsa takes the X_p as input, the other six variants have X_l as input. Figure 7.8 and Table 7.2 demonstrate the qualitative and quantitative comparisons. One can see that IDeudemon achieves overall better quantitative measures than its variants of high quality generation model. Specifically, BOPB, GPEN, RestoreFormer, and VQFR are weak in inpainting the irregular white gaps in X_l, BOPB alters the hue of the image, GPEN and RestoreFormer often suffer from artifacts at face contours, and VQFR sometimes produces blurry details (see the teeth). Although CodeFormer does a good job in filling in the white gaps, it tends to smooth out the whole faces and changes the clothing.

By discarding the visual similarity assistance, the results of w/o vsa cannot retain the identity-agnostic features. For instance, the background, hairstyle, accessories, and the clothing. Moreover, artifacts and unnatural splotches appear randomly, which affect the visual perception. Although w/o vsa performs slightly better in identity protection, its utility performance has deteriorated significantly. These imply that visual similarity assistance plays an import role in synthesizing realistic details and preserving utility.

It can be observed that only the complete loss function combination achieves the optimal results. It proves that L_{rec} reduces artifacts and preserves visual similarity, L_{adv} enhances realism, L_{comp} improves clarity in the eyes and mouth, and L_{id} maintains the protected identity. As for *idloss*, it can generate high quality faces; however, when applying the same protective perturbation, it generates face that looks more like the original face X than X_l. The privacy indicators of *idloss* demonstrate that our adjustment of original identity preserving loss can better protect the human identity.

Overall, IDeudemon shows superior performance to its variants, demonstrating the effectiveness of Step II's architecture and the adjusted identity preserving loss.

7.5 Discussion

We want to emphasize that while elements of IDeudemon are built on well-understood 3D reconstruction principles (dating back to Vetter and Blanz) and blind face restoration, our core contribution is new and essential. The key to making IDeudemon jump out of the annoying privacy–utility tradeoff is the "divide and conquer" idea that protects privacy and preserves utility in two sequential steps, the identity is protected at 3D space through a parametric NeRF model, both of which have not appeared previously in the literature. In addition, we pick the most suitable GAN model and perturbation range for our approach through sufficient experiments. We have also designed visual similarity assistance and adjusted the loss function so as to better finish the de-identification task.

7.6 Conclusion

In this chapter, we propose a novel two-step face de-identification method that conducts "divide and conquer" strategy to solve the challenging privacy–utility tradeoff problem. By introducing advanced 3D parametric face fitting and obfuscating the disentangled ID code, we hide the real identity and endow the whole model with good explainability. Equipped with the visual similarity assistance and generative prior embedded GAN, our model can produce photo-realistic de-identified faces, allowing us to adjust the protection level while keeping good image utility. Extensive experiments demonstrate the superior capability of IDeudemon in face de-identification, outperforming prior arts.

References

1. P. Agrawal, P. Narayanan, Person de-identification in videos. IEEE Trans. Circuits Syst. Video Technol. **21**(3), 299–310 (2011)
2. H. Hukkelås, R. Mester, F. Lindseth, Deepprivacy: a generative adversarial network for face anonymization, in *International Symposium on Visual Computing* (Springer, Berlin, 2019), pp. 565–578
3. M. Maximov, I. Elezi, L. Leal-Taixé, Ciagan: conditional identity anonymization generative adversarial networks, in *Proceedings of the IEEE/CVF Conference on Computer Vision and Pattern Recognition* (2020), pp. 5447–5456
4. T. Li, L. Lin, Anonymousnet: Natural face de-identification with measurable privacy, in *Proceedings of the IEEE Conference on Computer Vision and Pattern Recognition Workshops* (2019)
5. X. Gu, W. Luo, M.S. Ryoo, Y.J. Lee, Password-conditioned anonymization and deanonymization with face identity transformers, in *Computer Vision–ECCV 2020: 16th European Conference, Glasgow, UK, August 23–28, 2020, Proceedings, Part XXIII 16* (Springer, Berlin, 2020), pp. 727–743
6. J. Cao, B. Liu, Y. Wen, R. Xie, L. Song, Personalized and invertible face de-identification by disentangled identity information manipulation, in *Proceedings of the IEEE/CVF International Conference on Computer Vision* (2021), pp. 3334–3342
7. Y. Wen, B. Liu, M. Ding, R. Xie, L. Song, Identitydp: differential private identification protection for face images. Neurocomputing **501**, 197–211 (2022)
8. Y. Luo, J. Zhu, K. He, W. Chu, Y. Tai, C. Wang, J. Yan, Styleface: towards identity-disentangled face generation on megapixels, in *Computer Vision–ECCV 2022: 17th European Conference, Tel Aviv, Israel, October 23–27, 2022, Proceedings, Part XVI* (Springer, Berlin, 2022), pp. 297–312
9. L. Zhai, Q. Guo, X. Xie, L. Ma, Y. E. Wang, Y. Liu, A3gan: attribute-aware anonymization networks for face de-identification, in *Proceedings of the 30th ACM International Conference on Multimedia* (2022), pp. 5303–5313
10. S.J. Oh, R. Benenson, M. Fritz, B. Schiele, Faceless person recognition: privacy implications in social media, in *European Conference on Computer Vision* (Springer, Berlin, 2016), pp. 19–35
11. E.M. Newton, L. Sweeney, B. Malin, Preserving privacy by de-identifying face images. IEEE Trans. Knowl. Data Eng. **17**(2), 232–243 (2005)
12. R. Gross, L. Sweeney, F. De la Torre, S. Baker, Model-based face de-identification, in *2006 Conference on Computer Vision and Pattern Recognition Workshop (CVPRW'06)* (IEEE, Piscataway, 2006), pp. 161–161
13. A. Jourabloo, X. Yin, X. Liu, Attribute preserved face de-identification, in *2015 International Conference on Biometrics (ICB)* (IEEE, Piscataway, 2015), pp. 278–285
14. X. Yang, Y. Dong, T. Pang, H. Su, J. Zhu, Y. Chen, H. Xue, Towards face encryption by generating adversarial identity masks, in *Proceedings of the IEEE/CVF International Conference on Computer Vision* (2021), pp. 3897–3907
15. Y. Zhong, W. Deng, Opom: customized invisible cloak towards face privacy protection. IEEE Trans. Pattern Anal. Mach. Intell. **45**(3), 3590–3603
16. S. Hu, X. Liu, Y. Zhang, M. Li, L.Y. Zhang, H. Jin, L. Wu, Protecting facial privacy: generating adversarial identity masks via style-robust makeup transfer, in *Proceedings of the IEEE/CVF Conference on Computer Vision and Pattern Recognition* (2022), pp. 15014–15023

17. A.B. Arrieta, N. Díaz-Rodríguez, J. Del Ser, A. Bennetot, S. Tabik, A. Barbado, S. García, S. Gil-López, D. Molina, R. Benjamins et al., Explainable artificial intelligence (XAI): concepts, taxonomies, opportunities and challenges toward responsible AI, Inform. Fusion **58**, 82–115 (2020)
18. B. Gecer, S. Ploumpis, I. Kotsia, S. Zafeiriou, Ganfit: generative adversarial network fitting for high fidelity 3d face reconstruction, in *Proceedings of the IEEE/CVF Conference on Computer Vision and Pattern Recognition* (2019), pp. 1155–1164
19. A. Lattas, S. Moschoglou, S. Ploumpis, B. Gecer, A. Ghosh, S. Zafeiriou, Avatarme++: facial shape and brdf inference with photorealistic rendering-aware gans. IEEE Trans. Pattern Anal. Mach. Intell. **44**(12), 9269–9284 (2021)
20. L. Wang, Z. Chen, T. Yu, C. Ma, L. Li, Y. Liu, Faceverse: a fine-grained and detail-controllable 3d face morphable model from a hybrid dataset, in *Proceedings of the IEEE/CVF Conference on Computer Vision and Pattern Recognition* (2022), pp. 20333–20342
21. V. Blanz, T. Vetter, A morphable model for the synthesis of 3d faces, in *Proceedings of the 26th Annual Conference on Computer Graphics and Interactive Techniques* (1999), pp. 187–194
22. M. Sela, E. Richardson, R. Kimmel, Unrestricted facial geometry reconstruction using image-to-image translation, in *Proceedings of the IEEE International Conference on Computer Vision* (2017), pp. 1576–1585
23. T. Li, T. Bolkart, M.J. Black, H. Li, J. Romero, Learning a model of facial shape and expression from 4d scans. ACM Trans. Graph. **36**(6), 194–1 (2017)
24. B. Mildenhall, P.P. Srinivasan, M. Tancik, J.T. Barron, R. Ramamoorthi, R. Ng, Nerf: representing scenes as neural radiance fields for view synthesis. Commun. ACM **65**(1), 99–106 (2021)
25. A. Yu, V. Ye, M. Tancik, A. Kanazawa, pixelnerf: neural radiance fields from one or few images, in *Proceedings of the IEEE/CVF Conference on Computer Vision and Pattern Recognition* (2021), pp. 4578–4587
26. M. Oechsle, S. Peng, A. Geiger, Unisurf: unifying neural implicit surfaces and radiance fields for multi-view reconstruction, in *Proceedings of the IEEE/CVF International Conference on Computer Vision* (2021), pp. 5589–5599
27. P. Rao, B. Mallikarjun, G. Fox, T. Weyrich, B. Bickel, H. Pfister, W. Matusik, A. Tewari, C. Theobalt, M. Elgharib, Vorf: volumetric relightable faces (2022)
28. D. Wang, P. Chandran, G. Zoss, D. Bradley, P. Gotardo, Morf: morphable radiance fields for multiview neural head modeling, in *ACM SIGGRAPH 2022 Conference Proceedings* (2022), pp. 1–9
29. S. Galanakis, B. Gecer, A. Lattas, S. Zafeiriou, 3dmm-rf: convolutional radiance fields for 3d face modeling, in *Proceedings of the IEEE/CVF Winter Conference on Applications of Computer Vision* (2023), pp. 3536–3547
30. Y. Hong, B. Peng, H. Xiao, L. Liu, J. Zhang, Headnerf: a real-time nerf-based parametric head model, in *Proceedings of the IEEE/CVF Conference on Computer Vision and Pattern Recognition* (2022), pp. 20374–20384
31. X. Wang, Y. Li, H. Zhang, Y. Shan, Towards real-world blind face restoration with generative facial prior, in *Proceedings of the IEEE/CVF Conference on Computer Vision and Pattern Recognition* (2021), pp. 9168–9178
32. C. Chen, X. Li, L. Yang, X. Lin, L. Zhang, K.-Y. K. Wong, Progressive semantic-aware style transformation for blind face restoration, in *Proceedings of the IEEE/CVF Conference on Computer Vision and Pattern Recognition* (2021), pp. 11896–11905
33. Y. Chen, Y. Tai, X. Liu, C. Shen, J. Yang, Fsrnet: end-to-end learning face super-resolution with facial priors, in *Proceedings of the IEEE Conference on Computer Vision and Pattern Recognition* (2018), pp. 2492–2501
34. X. Li, C. Chen, S. Zhou, X. Lin, W. Zuo, L. Zhang, Blind face restoration via deep multi-scale component dictionaries, in *Computer Vision–ECCV 2020: 16th European Conference, Glasgow, UK, August 23–28, 2020, Proceedings, Part IX 16* (Springer, Berlin, 2020), pp. 399–415

35. Z. Wang, J. Zhang, R. Chen, W. Wang, P. Luo, Restoreformer: high-quality blind face restoration from undegraded key-value pairs, in *Proceedings of the IEEE/CVF Conference on Computer Vision and Pattern Recognition* (2022), pp. 17512–17521
36. Y. Gu, X. Wang, L. Xie, C. Dong, G. Li, Y. Shan, M.-M. Cheng, Vqfr: blind face restoration with vector-quantized dictionary and parallel decoder, in *Computer Vision–ECCV 2022: 17th European Conference, Tel Aviv, Israel, October 23–27, 2022, Proceedings, Part XVIII* (Springer, Berlin, 2022), pp. 126–143
37. S. Zhou, K.C. Chan, C. Li, C.C. Loy, Towards robust blind face restoration with codebook lookup transformer (2022). arXiv preprint arXiv:2206.11253
38. J. Gu, Y. Shen, B. Zhou, Image processing using multi-code gan prior, in *Proceedings of the IEEE/CVF Conference on Computer Vision and Pattern Recognition* (2020), pp. 3012–3021
39. S. Menon, A. Damian, S. Hu, N. Ravi, C. Rudin, Pulse: self-supervised photo upsampling via latent space exploration of generative models, in *Proceedings of the IEEE/CVF Conference on Computer Vision and Pattern Recognition* (2020), pp. 2437–2445
40. Z. Wan, B. Zhang, D. Chen, P. Zhang, D. Chen, J. Liao, F. Wen, Bringing old photos back to life, in *Proceedings of the IEEE/CVF Conference on Computer Vision and Pattern Recognition* (2020), pp. 2747–2757
41. T. Yang, P. Ren, X. Xie, L. Zhang, Gan prior embedded network for blind face restoration in the wild, in *Proceedings of the IEEE/CVF Conference on Computer Vision and Pattern Recognition* (2021), pp. 672–681
42. M.R. Koujan, M.C. Doukas, A. Roussos, S. Zafeiriou, Head2head: video-based neural head synthesis, in *2020 15th IEEE International Conference on Automatic Face and Gesture Recognition (FG 2020)* (IEEE, Piscataway, 2020), pp. 16–23
43. Y. Guo, L. Cai, J. Zhang, 3d face from x: learning face shape from diverse sources. IEEE Trans. Image Process. **30**, 3815–3827 (2021)
44. L. Tran, X. Liu, Nonlinear 3d face morphable model, in *Proceedings of the IEEE Conference on Computer Vision and Pattern Recognition* (2018), pp. 7346–7355
45. Y. Deng, J. Yang, S. Xu, D. Chen, Y. Jia, X. Tong, Accurate 3d face reconstruction with weakly-supervised learning: from single image to image set, in *Proceedings of the IEEE/CVF Conference on Computer Vision and Pattern Recognition Workshops* (2019)
46. C. Yu, J. Wang, C. Peng, C. Gao, G. Yu, N. Sang, Bisenet: bilateral segmentation network for real-time semantic segmentation, in *Proceedings of the European Conference on Computer Vision (ECCV)* (2018), pp. 325–341
47. C. Yu, C. Gao, J. Wang, G. Yu, C. Shen, N. Sang, Bisenet v2: bilateral network with guided aggregation for real-time semantic segmentation. Int. J. Comput. Vis. **129**, 3051–3068 (2021)
48. T. Karras, S. Laine, M. Aittala, J. Hellsten, J. Lehtinen, T. Aila, Analyzing and improving the image quality of stylegan, in *Proceedings of the IEEE/CVF Conference on Computer Vision and Pattern Recognition* (2020), pp. 8110–8119
49. O. Ronneberger, P. Fischer, T. Brox, U-net: convolutional networks for biomedical image segmentation, in *Medical Image Computing and Computer-Assisted Intervention–MICCAI 2015: 18th International Conference, Munich, Germany, October 5–9, 2015, Proceedings, Part III 18* (Springer, Berlin, 2015), pp. 234–241
50. J. Johnson, A. Alahi, L. Fei-Fei, Perceptual losses for real-time style transfer and super-resolution, in *European Conference on Computer Vision* (Springer, Berlin, 2016), pp. 694–711
51. C. Ledig, L. Theis, F. Huszár, J. Caballero, A. Cunningham, A. Acosta, A. Aitken, A. Tejani, J. Totz, Z. Wang et al., Photo-realistic single image super-resolution using a generative adversarial network, in *Proceedings of the IEEE Conference on Computer Vision and Pattern Recognition* (2017), pp. 4681–4690
52. K. Simonyan, A. Zisserman, Very deep convolutional networks for large-scale image recognition (2014). arXiv preprint arXiv:1409.1556
53. I. Goodfellow, J. Pouget-Abadie, M. Mirza, B. Xu, D. Warde-Farley, S. Ozair, A. Courville, Y. Bengio, Generative adversarial networks. Commun. ACM **63**(11), 139–144 (2020)

54. T.-C. Wang, M.-Y. Liu, J.-Y. Zhu, A. Tao, J. Kautz, B. Catanzaro, High-resolution image synthesis and semantic manipulation with conditional gans, in *Proceedings of the IEEE Conference on Computer Vision and Pattern Recognition* (2018), pp. 8798–8807
55. K. He, G. Gkioxari, P. Dollár, R. Girshick, Mask R-CNN, in *Proceedings of the IEEE International Conference on Computer Vision* (2017), pp. 2961–2969
56. L.A. Gatys, A.S. Ecker, M. Bethge, Image style transfer using convolutional neural networks, in *Proceedings of the IEEE Conference on Computer Vision and Pattern Recognition* (2016), pp. 2414–2423
57. F. Schroff, D. Kalenichenko, J. Philbin, Facenet: a unified embedding for face recognition and clustering, in *Proceedings of the IEEE Conference on Computer Vision and Pattern Recognition* (2015), pp. 815–823
58. J. Deng, J. Guo, N. Xue, S. Zafeiriou, Arcface: additive angular margin loss for deep face recognition, in *Proceedings of the IEEE Conference on Computer Vision and Pattern Recognition* (2019), pp. 4690–4699
59. T. Karras, S. Laine, T. Aila, A style-based generator architecture for generative adversarial networks, in *Proceedings of the IEEE/CVF Conference on Computer Vision and Pattern Recognition* (2019), pp. 4401–4410
60. T. Karras, T. Aila, S. Laine, J. Lehtinen, Progressive growing of gans for improved quality, stability, and variation (2017). arXiv preprint arXiv:1710.10196
61. D. Yi, Z. Lei, S. Liao, S. Z. Li, Learning face representation from scratch (2014). arXiv preprint arXiv:1411.7923
62. Q. Cao, L. Shen, W. Xie, O. M. Parkhi, A. Zisserman, Vggface2: a dataset for recognising faces across pose and age, in *2018 13th IEEE International Conference on Automatic Face & Gesture Recognition (FG 2018)* (IEEE, Piscataway, 2018), pp. 67–74
63. C.-Y. Wu, Q. Xu, U. Neumann, Synergy between 3dmm and 3d landmarks for accurate 3d facial geometry, in *2021 International Conference on 3D Vision (3DV)* (2021)
64. Y. Feng, H. Feng, M. J. Black, T. Bolkart, Learning an animatable detailed 3d face model from in-the-wild images. ACM Trans. Graph. **40**(4), 1–13 (2021)
65. D.P. Kingma, J. Ba, Adam: a method for stochastic optimization (2014). arXiv preprint arXiv:1412.6980

Chapter 8
Deep Motion Flow Guided Reversible Face Video De-identification

8.1 Introduction

The proliferation of smartphones and short-video platforms has changed the way people create and consume video. Ordinary individuals have become the primary producers and consumers of video activities [1]. With the surge in the number of online videos, the sensitive information (such as human faces) contained in these videos has caused unprecedented violations in the field of personal privacy protection [2]. New privacy laws and regulations begin to forbid the public disclosure of personal sensitive information. However, since the access and utilization of such videos are neither easy to monitor nor to prevent, it is essential to grant users the option to obfuscate themselves out of these videos.

Advanced computer vision technology and blooming online social networks have greatly facilitated both daily social interactions and face videos sharing [3]. While the media users are willing to guard their personal privacy, they are also eager to enjoy the convenience of advanced identity-agnostic computer vision applications. These applications do not need to identify the people in the videos, for instance, face detection, face reenactment, emotion analysis, action recognition, and so on. Therefore, maintaining the utility of identity-protected videos to support existing identity-agnostic tasks and normal online social use becomes a new and appealing topic. In addition, the Internet is not an extrajudicial land. When an incident such as a crime occurs, authorities should be able to examine the original videos for forensics purposes.

Reversible face video de-identification is an effective solution to the aforementioned issues. But it is very challenging to design a satisfactory technique to achieve this target. On the one hand, it requires obfuscating the sensitive

Main contents of this chapter have been published in "Wen, Y., Liu, B., Cao, J., Xie, R., Song, L., & Li, Z. (2022). IdentityMask: deep motion flow guided reversible face video de-identification. IEEE Transactions on Circuits and Systems for Video Technology, 32(12), 8353–8367."

Y. Wen et al., *Face De-identification: Safeguarding Identities in the Digital Era*,
https://doi.org/10.1007/978-3-031-58222-6_8

identity information of the subject while minimizing distortion or changes in other nonidentity features [4–6], i.e., ensuring visual similarity including appearance, posture, expression, background information, etc. On the other hand, in the case of "after-the-fact forensics," it allows the authorized party to fully restore the anonymous videos [7].

Existing face video privacy-preserving methods [4–6, 8–12] only focus on the former aspect and lack the restoration ability, which does not meet the privacy requirements of keeping pace with the times. Furthermore, these methods process video frame by frame without considering the temporal relationship between frames. This can easily make the de-identified video flicker due to temporal inconsistency and cause excessive computational overhead.

In order to overcome the above problems, in this chapter, we present a novel and effective reversible face video de-identification modular framework guided by deep motion flow, called IdentityMask. Our framework contains two main functional modules (*Protection Module* and *Recovery Module*), both of which are guided by the crucial *Motion Flow Module*, while an *Affine Transformation Module* provides simple but reliable assistance. Instead of per-frame processing, it lets only the first affined frame go through the *Protection/Recovery Module* and calculates the deep motion flow between every two adjacent frames via a motion flow generator. Then the subsequent de-identified/recovered frames can be generated based on the first protected/recovered frame by the guide of the relative motion representation. All the synthesized videos can be visually pleasing without flickering. Also, we design a discrete key space where keys condition identity changes to securely enable the recovery transformation only for the authorized parties. Specifically, any video that the user wants to obfuscate will be transformed into the de-identified one with an assigned Ukey (a number that matches the user's UID). Then, given a de-identified video, the original video can only be recovered if the correct key and the trained recovery pipeline are provided. We further increase security as follows: Given an anonymized video, even the trained recovery pipeline is stolen, if a wrong key is provided, it changes to a new identity that is still different from the original one (Fig. 8.1, "Wrong Key"), with a natural appearance. When the framework is used in practical applications, the Ukey can be a number defined according to the specific situation. For example, specified by the user, distributed by the video platform, and so on.

This chapter is built upon our prior work [13] and [14] with multiple improvements. Compared with [13], we add a recovery process and achieve reversible face video de-identification, which is more conducive to the establishment of orderly online social networks. We also add Ukey to enable users to control the de-identification process. The method in [14] works on still images and cannot be directly applied to videos, while in this chapter, we use two specifically designed modules to process videos. In addition, the reference identity in [14] is obtained by randomly selecting k (during the experiment, we set $k = 3000$) different identities from the training set, which is inconvenient and may cause legal disputes. We solve

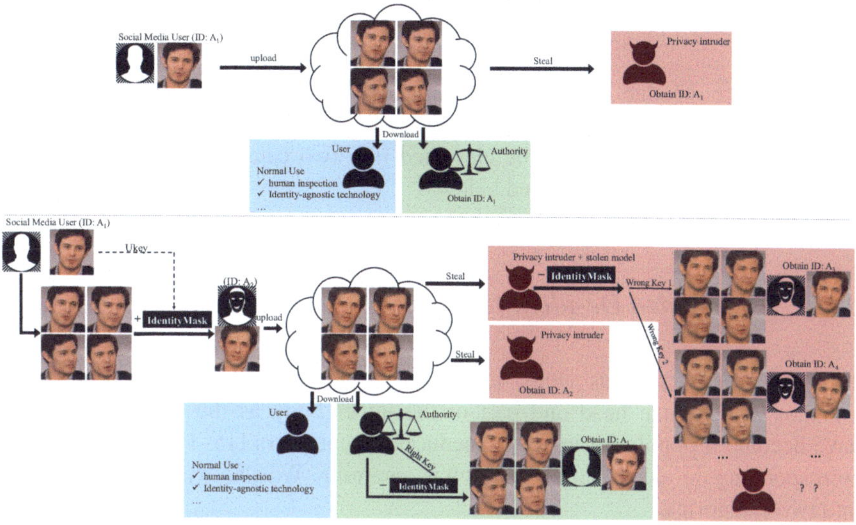

Fig. 8.1 Comparison between a vulnerable social media platform (left panel) and a IdentityMask protected social media platform (right panel) in maintaining normal use, safeguarding legal supervision as well as handling malicious privacy intruders for stealing personal identity information

this shortcoming by leveraging random seeds conditioned on Ukeys to generate reference identities, which is more flexible and gets rid of the need for auxiliary faces.

In summary, the main contributions of this chapter are described as follows:

- To the best of our knowledge, the proposed IdentityMask is the first method that can conduct reversible de-identification for face videos. Our proactive defense technique well addresses the growing concerns about personal privacy protection during online video sharing. On the one hand, the users can protect individual identity with a certain Ukey (equivalent to a password) before sharing. On the other hand, when the identity-protected video is released, the authorized party can still obtain the recovered video with the original identity through the correct Ukey, while it is difficult for unauthorized parties to infer the true identity.
- We introduce deep motion flow into video de-identification tasks to avoid per-frame processing. We show that the *Motion Flow Module* can provide important guidance for IdentityMask pipeline to generate identity-protected/identity-recovered videos, resulting in significantly improved synthesis quality and reduced computational overhead.
- Experimental results on a diverse face video dataset (gender, ethnicity, age, etc.) have demonstrated the effectiveness of our proposed IdentityMask. In addition, we introduce evaluation metrics designed for videos, which are lacking in existing literature.

8.2 Related Work

To our best knowledge, our work is unique and there is no previous similar work to directly compare with. Nevertheless, it is closely related to previous video de-identification work, which can be classified into two categories according to the application scenarios, as described below.

8.2.1 Face Video De-identification

The face videos, such as vlogs, live-streaming sales, speeches, and interviews, are shot with human head and part of the upper body as the main subject and have become a popularity in social media in recent years [15–17]. Therefore, the corresponding de-identification research is emerging. We classify these approaches into two categories.

Identity-Swapping-Based Methods Replacing the identity in a face video with someone else is a straightforward but effective idea of de-identification. The "someone" here can be either a real identity provider or a somehow synthesized identity that does not exist in reality. Generally, the latter is a more thorough way of privacy protection.

Zhu et al. [12] applied deepfake technology to de-identify medical examination videos by explicitly swapping the patients' faces with open-source characters. However, such simple operation will lead to an extreme deterioration in visual similarity, and thus more skillful identity-swapping-based methods are proposed. With several pretrained active appearance models (AAMs), Samarzija et al. [9] found the best fit model of the original face and swapped the face region with another face taken from the training dataset. Meden et al. [10] replaced the original faces with surrogates generated from a small number of identities. Instead of synthesizing the surrogate faces through simple pixel averaging, they used a convolutional neural network (CNN) to generate artificial surrogates. Li et al. [11] used a trained facial attribute transfer model (FATM) to map the nonidentity-related facial attributes to the face of donors, who were a small number (usually 2–3) of consented subjects. Gafni et al. [5] utilized a multilevel face descriptor to convert the identity of the original face to that of the target face. Specifically, the removal of identity was done via distancing the face descriptors of the output video from those of the original image. Maximov et al. [6] removed the identification characteristics of input people in the bottleneck of the generator via a one-hot label that encoded the desired identity; meanwhile, they leveraged the input landmark images with some original identity information left to preserve the pose; thus the generated identity was a composition of both the landmark identity and the desired identity.

Identity Disentanglement-Based Methods Although the former kind of methods have evolved to a stage with amazing results, its reliance on auxiliary identities can make it difficult to apply under increasingly stringent regulations. For example, consent from the target identity provider should be obtained regularly, which is kind of inconvenient.

Consequently, another pattern that deals with face video de-identification through certain face models by training to extract facial feature representations begins to rise. Once the representations have been disentangled, a de-identified face video can then be generated based on the new representations originated in which the protected identity information has been eliminated, reduced, or obfuscated. During this time, a new virtual identity will generate. Our method follows this pattern.

Gross et al. [8] factorized input images into identity and nonidentity factors using a generative multifactor model and then applied a de-identification algorithm on the combined factorized data before using the bases of the multifactor model to reconstruct de-identified images. With the development of deep neural network (DNN), deep face models can better undertake the task of disentanglement. Ren et al. [4] employed a multitask extension of the generative adversarial network (GAN), where a face anonymizer tried to minimize the identification accuracy and an activity detector tried to maximize spatial action detection performance.

We provide a comprehensive comparison between the previous face video de-identification methods and ours in Table 8.1.

Table 8.1 A comparison to the existing face video de-identification methods

	Gross [8]	Samarzija [9]	Meden [10]	Li [11]	Zhu [12]	Ren [4]	Gafni [5]	Maximov [6]	Ours
Without auxiliary faces	Yes	No	No	No	No	Yes	No	No	Yes
Demonstrated on a diverse video dataset (gender, ethnicity, age, etc.)	No	No	Yes	No	No	Yes	Yes	Yes	Yes
Demonstrated on a diverse face video dataset (gender, ethnicity, age, etc.)	No	No	No	No	No	No	Yes	No	Yes
Without per-frame processing	No	No	No	No	No	No	No	No	Yes
Recover original face	No	No	No	No	No	No	No	No	Yes
Reference to a comparison with ours						Fig. 8.5	Fig. 8.6	Fig. 8.5	

8.2.2 Surveillance Video De-identification

Video surveillance systems have been omnipresent for a considerable time, with large systems being deployed in strategic places such as public transportation, airports, city centers, or residential areas. In order to address the never-ending concerns about personal privacy protection, a large amount of targeted de-identification technologies have been proposed. As the surveillance videos typically contain multiple people (with full body) and complex surrounding environment, these methods always attach great importance to efficient face detection and tracking and apply anonymization on the segmented origins. Here we classify them into three categories.

Obfuscation-Based Methods These methods achieve video de-identification by obfuscating each frame's privacy-sensitive region in some way. Specifically, Dufaux et al. [18] used domain scrambling methods to achieve distortion. Schiff et al. [19] employed solid ellipsoidal overlays, while minimized the overlay area to maximize the remaining observable region of the scene. Chen et al. [20] implemented an EMHI approach to obscure the entire body. Agrawal et al. [21] applied the exponential blur of pixels in the voxel or line integral convolution. Mrityunjay et al. [22] obscured the segmented bounding box region by using Gaussian Blur of the pixels and binarizing the intensity values. Ivasic-Kos et al. [23] applied 2D Gaussian filtering to automatically obfuscate the human body shape information. Blažević et al. [24] replaced humans with rendered 3D human models. Ryoo et al. [25] presented an inverse super-resolution (ISR) paradigm that used extreme low-resolution (e.g., 16×12) videos to achieve de-identification and benefit activity recognition. Flouty et al. [26] introduced a sliding window smoother for temporal smoothing on the detections. [27] obfuscated the privacy-sensitive parts at multiple privacy levels by using a random corruption matrix. Kim et al. [28] fundamentally protected privacy by blurring unwanted blocks in images, yet ensured that the robots could understand the video for their perception. Wang et al. [29] used a lensless coded aperture (CA) camera, which placed only a coded aperture in front of an image sensor, and the resulting CA images would be visually unrecognizable and were difficult to restore with high fidelity. Zhou et al. [1] proposed a novel PsOP framework that was extendable to any potential privacy-sensitive objects pixelation after leveraging pretrained detection networks as the backbone. Tu et al. [30] generated bounding boxes to cover the regions of interest, and then the pixels inside bounding boxes could be modified to achieve a certain degree of content-obscuring to obscure the person-identifiable contents.

Style Transfer-Based Methods Style transfer has also been used to do de-identification. Winkler et al. [31] generated an abstracted version of the security regions that showing only outlines of persons. Erdélyi et al. [32] presented a resource-aware cartooning privacy protection filter that converted raw images into abstracted frames where the privacy revealing details were removed. Brkić et al. [33] altered the appearance of the segmented pedestrians through a neural art algorithm that used the responses of a DNN to render the pedestrian images in a different style. [34] proposed two privacy protection schemes by using false colors on entire images. PECAM [7] converted the real-world images (domain-X) into the privacy-enhanced ones (domain-Y) through cartoon style rendering.

Identity Disentanglement-Based Methods Recently, the development of deep CNNs has also inspired new methods based on identity disentanglement [35]. Li et al. [36] developed an encoder–decoder network architecture that could separately disentangle the facial feature representation into an appearance code and an identification code. The anonymous face was synthesized by recombining the original identity code and another appearance code from the target set to protect the individual privacy. Proença et al. [37] used a binary vector labelling ID, gender, ethnicity, age, and hairstyle predicted by an attribute classifier to keep full control over the appearance of the anonymous faces.

Especially, among all these video surveillance de-identification methods, there exist five methods [7, 24, 27, 34, 37] that are reversible and can recover the original scene. Therefore, it is imperative to develop similar reversible de-identification technology for face videos. These five technologies focus on the accurate recording of events in supervised scenes, while little attention is paid to the generation of subtle details due to the original low resolution. In contrast, we strive to generate visual-pleasing facial details and maintain accurate facial motion.

8.3 Preliminaries of Problem Formulation

A reversible face video de-identification model generally can be viewed as a combination of a complex function δ and its inverse function δ^{-1}. To be more specific, the function δ maps a given face video $V = (v_1, v_2, \cdots, v_n)$ (v_i represents the ith frame) to a de-identified video $V' = (v'_1, v'_2, \cdots, v'_n)$, aiming to conceal the real identity, and can be formulated as

$$\begin{aligned} \delta(V) &= V' \\ s.t.: 1 \leq i \leq n&, \mathrm{ID}\{v_i\} \neq \mathrm{ID}\{v'_i\}. \end{aligned} \tag{8.1}$$

After this, video V' can still be used normally, and when given the right key, the function δ^{-1} can restore a video $V_r = (v_{r,1}, v_{r,2}, \cdots, v_{r,n})$ with the original identity, but if the key is wrong, the function δ^{-1} restores a visual-pleasing video $V_w = (v_{w,1}, v_{w,2}, \cdots, v_{w,n})$ whose identity is different from the original video's identity. It can be formulated as follows: When the right key is given:

$$\begin{aligned} \delta^{-1}(V') &= V_r \\ s.t.: 1 \leq i \leq n&, \mathrm{ID}\{v_{r,i}\} = \mathrm{ID}\{v_i\}; \end{aligned} \tag{8.2}$$

and when the wrong key is given:

$$\delta^{-1}(V') = V_w \tag{8.3}$$
$$s.t.: 1 \leq i \leq n, \mathrm{ID}\{v_{w,i}\} \neq \mathrm{ID}\{v_i\}.$$

8.4 Deep Motion Flow Guided Reversible Face Video De-identification

In this section, we propose a modular architecture, called IdentityMask, to address the reversible face video de-identification problem. From the perspective of realized function, IdentityMask includes two-directional mappings: a de-identification process and a recovery process. When given an original nonprotected face video V, the de-identification process aims to transform it into an identity-protected one (V'), whose identity change conditioned on the Ukey, and the recovery process aims to transform the de-identified video V' into an identity-recovered one (V_r with right key or V_w with wrong key). Figure 8.2 illustrates the whole pipeline.

From the perspective of framework structure, IdentityMask consists of two main functional modules: the identity protection module *Protection Module* and the identity restoration module *Recovery Module*, both of which are guided by the vital *Motion Flow Module*. With the simple but reliable assistance of the

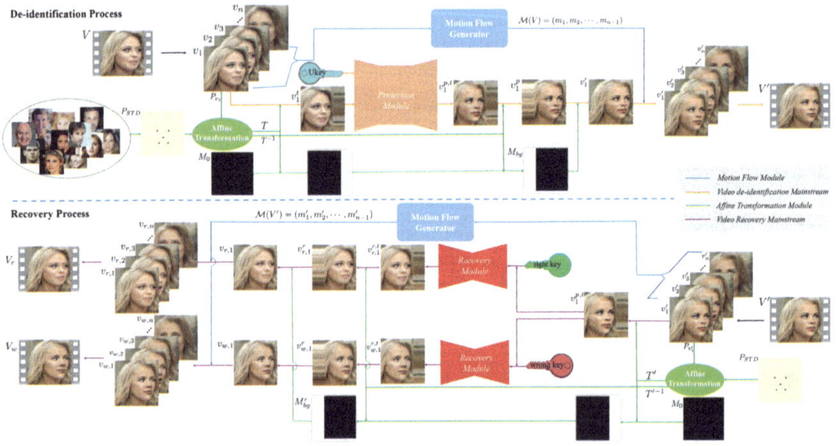

Fig. 8.2 The overall architecture of the proposed reversible face video de-identification method, IdentityMask. Our framework consists of two processes: The de-identification process provides a protective mask for identity information, while the recovery process removes the protective mask if and only if the right key is provided. The former relies on the *Protection Module*, and the latter relies on the *Recovery Module*, both of which are guided by the crucial *Motion Flow Module*, and are assisted by the simple but reliable *Affine Transformation Module*

Table 8.2 Notations

Superscript	t	Affined frame
	p	ID-protected frame
	′	ID-protected frame with background
	r	ID-recovered frame
Subscript	r	ID right recovered frame
	w	ID wrong recovered frame
Model	$\mathcal{M}$	Motion flow generator
	$\mathcal{F}$	Fusion network

Affine Transformation Module, IdentityMask efficiently achieves reversible de-identification. In the following subsections, we first introduce the four modules, respectively, and then describe the entire IdentityMask pipeline. Notations used in this section are summarized in Table 8.2.

8.4.1 Protection Module

We achieve de-identification by following the identity disentanglement pattern. As shown in Fig. 8.4, when given an original clean face frame v_1^t, we apply an identity encoder and an attribute encoder to extract two disentangled representations of the latent space, denoted as $r_{id}(v_1^t)$ and $r_{attr}(v_1^t)$. Among them, the identity representation r_{id} contains all the information relevant to the identity that affects face verification systems to judge whether it is the same person, and the attribute representation r_{attr} contains the rest of information carried by the image that guarantees the visual similarity (e.g., pose, expression, overall structure, background, and so on). Based on this, we firstly use the Ukey as a randomness seed to generate a reference identity vector r_{refer} whose size equals to $r_{id}(v_1^t)$, which is formulated as

$$r_{refer} = \mathcal{R}_{Ukey}. \tag{8.4}$$

Here the Ukey is a number that uniquely represents the user's identity. Then, a component vector $r_{\perp}(v_1^t)$ that is orthogonal to $r_{id}(v_1^t)$ in r_{refer} can be decomposed as follows:

$$r_{\perp}(v_1^t) = r_{refer} - (r_{id}(v_1^t) \cdot r_{refer}) \cdot r_{id}(v_1^t). \tag{8.5}$$

It allows us to create a new identity $r_{new}(v_1^t)$ by rotating $r_{id}(v_1^t)$ with a controllable parameter θ, and we denote it as

$$r_{new} = r_{id}(v_1^t) \cdot \cos\theta + r_{\perp}(v_1^t) \cdot \sin\theta. \tag{8.6}$$

Finally we synthesize the de-identified face $v_1^{p,t}$ with new identity representation r_{new} and original attribute representation $r_{attr}(v_1^t)$ through a well-trained fusion network as follows:

$$v_1^{p,t} = \mathcal{F}(r_{new}, r_{attr}(v_1^t)). \tag{8.7}$$

8.4.2 Recovery Module

Given a de-identified face frame $v_1^{p,t}$, our *Recovery Module*, which is based on the same identity disentanglement network structure as the *Protection Module*, can restore the original frame with real identity if and only if the right key is provided. To be more specific, we firstly imply the aforementioned identity and attribute encoders to extract its identity representation $r_{id}(v_1^{p,t})$ and attribute representation $r_{attr}(v_1^{p,t})$, which has the relationship as

$$\begin{aligned} r_{id}(v_1^{p,t}) &= r_{new}, \\ r_{attr}(v_1^{p,t}) &= r_{attr}(v_1). \end{aligned} \tag{8.8}$$

Then when given the right key (i.e., R_key, which we define to be equal to the Ukey), the recovered identity embedding r_{rid} can be calculated as

$$r_{rid} = \frac{r_{id}(v_1^{p,t}) - \mathcal{R}_{R\_key} \cdot \sin\theta}{\cos\theta - A \cdot \sin\theta}, \tag{8.9}$$

where

$$A = \frac{\cos^2\theta - (r_{id}(v_{1,p}) - \mathcal{R}_{R\_key} \cdot \sin\theta) \cdot r_{id}(v_1^{p,t})}{\sin\theta \cdot \cos\theta}. \tag{8.10}$$

In fact, middle parameter A equals $r'_{id} \cdot r_{refer}$. Finally, the right recovered image with original real identity can be obtained as

$$v_{r,1}^{r,t} = \mathcal{F}(r'_{id}, r_{attr}(v_{1,p})). \tag{8.11}$$

8.4.3 Motion Flow Module

The above *Protection* and *Recovery Modules* work well for images, and it is straightforward to directly apply them to videos in a frame by frame way. However, since both modules rely on the disentanglement of latent convolutional features, direct per-frame processing is time-consuming. Typically, the flow estimation and

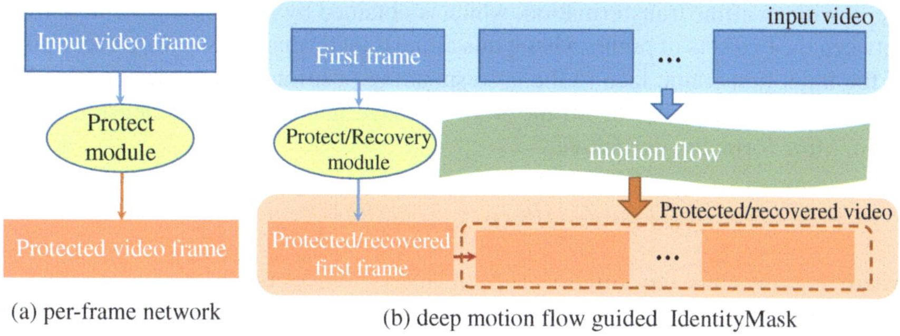

Fig. 8.3 Illustration of (**a**) existing face video de-identification technologies using per-frame network generation and (**b**) the proposed deep motion flow guided reversible face video de-identification

feature propagation are much faster than the computation of convolutional features [38], and consecutive face video frames are highly similar, so we exploit the similarity to reduce computational cost and achieve speedup. Specifically, either the *Protection Module* or the *Recovery Module* only processes the first frame; then we use a motion flow generator to calculate the relative motion flow of every two adjacent frames (see Fig. 8.3), which is denoted as

$$\mathcal{M} = (m_1, m_2, \cdots, m_{n-1}), \tag{8.12}$$

where m_i $(i \leq n)$ denotes the relative motion flow that can warp the processed (i.e., de-identified or recovered) ith frame to the next $(i+1)$th frame.

8.4.4 Affine Transformation Module

The position and pose of faces in online sharing videos vary widely, which usually differ from the "standard" frontal alignment that commonly used in large face datasets. However, it is well-known that computing deep representations by using a pretrained CNN does have a restriction: The test image needs to lie close to the image distribution trained by the CNN. Otherwise, the latent optimization may fail to reproduce on the test image, leading to poor feature maps. Therefore, directly applying *Protection Module* or *Recovery Module* on the first frame is invalid, and we design an affine transformation, which can standardize and restore the distribution of the first frame.

To be specific, we calculate several keypoints of all faces in the training datasets of *Protection* and *Recovery Module*, compute the average, and set it as the standard pattern (denoted as P_{STD}). Every time before the first frame is input to the *Protection* or *Recovery Module*, its keypoints (denoted as P_{v_1} or $P_{v_1'}$) are firstly computed in the same way. Then these keypoints are matched to the standard keypoint pattern

P_{STD} with an affine transformation, which is obtained by minimizing the distortion between the two sets of points. Using this affine transformation, we warp every pixel of the input first frame face to the corresponding position of the average face. We then copy the edge color to fill the warped image into the same dimension as the input. More formally, we denote

$$T = P_{v_1} \mho P_{STD}, \quad T' = P_{v_1'} \mho P_{STD}, \tag{8.13}$$

where T represents the affine transform matrix and $\mho$ denotes the affine transformation between two point patterns. Besides, we also need the inverse affine transformation to restore the original face position, and it is formulated as

$$T^{-1} = P_{STD} \mho P_{v_1}, \quad T'^{-1} = P_{STD} \mho P_{v_1'}, \tag{8.14}$$

where T^{-1} represents the inverse transform matrix.

8.4.5 *The Entire IdentityMask Pipeline*

Our pipeline consists of a de-identification process and a recovery process (see Fig. 8.2).

The de-identification process takes the original clean video $V = (v_1, v_2, \cdots, v_n)$ as input. First of all, it is sent into the *Motion Flow Module*, where the motion flow generator generates the relative motion flow between every two adjacent frames, which is formulated as

$$\mathcal{M}(V) = (m_1, m_2, \cdots, m_{n-1}). \tag{8.15}$$

Based on this, the first frame v_1 firstly enters the *Affine Transformation Module* to generate the affine transform matrix T and the inverse affine transform matrix T^{-1} (see Eqs. (8.13) and (8.14)). Then the image v_1^t that lies in the "standard" distribution is obtained via

$$v_1^t = v_1 \cdot T. \tag{8.16}$$

This warped first frame v_1^t is sent to the *Protection Module*, through which the real identity is concealed and a new identity r_{new} conditioned on the Ukey is generated. We denote

$$v_1^{p,t} = \mathcal{F}(r_{new}, r_{attr}(v_1^t)). \tag{8.17}$$

Next, the de-identified frame $v_1^{p,t}$ restores to the same layout as the original input v_1 through an inverse affine transformation:

$$v_1^p = v_1^{p,t} \cdot T^{-1}. \tag{8.18}$$

In order to preserve the original background, a background mask M_{bg} is generated by applying a black image M_0 whose dimension is the same as the input v_1 through two affine transformations, which is denoted as

$$M_{bg} = M_0 \cdot T \cdot T^{-1}. \tag{8.19}$$

With the help of M_{bg}, we can get the de-identified first frame:

$$v_1' = v_{p,1} \cdot (1 - M_{bg}) + v_1 \cdot M_{bg}. \tag{8.20}$$

Finally, we can obtain the entire identity-protected video $V' = (v_1', v_2', \cdots, v_n')$ on the basis of the successfully de-identified first frame v_1' and the relative motion flow $\mathcal{M}(V)$. Specifically, for $1 < i \leq n$:

$$v_i' = v_{i-1}' \circledast m_{i-1}, \tag{8.21}$$

where $\circledast$ denotes the inference of v_i' with the former de-identified frame v_{i-1}' and the relative motion flow m_{i-1}.

The de-identification process is summarized in Algorithm 2. The recovery process is similar except that the *Protection Module* is replaced by the *Recovery Module* and is summarized in Algorithm 3.

Algorithm 2 De-identification process

Input: Original non-protected video $V = \{v_i\}_{i=1}^n$, Ukey.
Output: De-identified video $V' = \left\{v_i'\right\}_{i=1}^n$.

Generate the relative motion flow $\mathcal{M}(V) = \{m_i\}_{i=1}^{n-1}$
Generate affine transform matrixes T in Eq. (8.13) and T^{-1} in Eq. (8.14).
Generate background mask M_{bg} with black image M_0:
$$M_{bg} = M_0 \cdot T \cdot T^{-1}.$$
while $i = 1$ **do**
  $v_1^t = v_1 \cdot T.$
  Generate new identity embedding r_{new} with Ukey in Eq. (8.6) and de-identify the affined first frame:
$$v_{p,1}^t = \mathcal{F}(r_{new}, r_{attr}(v_1^t)).$$
  Restore the original layout: $v_1^p = v_1^{p,t} \cdot T^{-1}$.
  Generate de-identified first frame with preserved background:
$$v_1' = v_1^p \cdot (1 - M_{bg}) + v_1 \cdot M_{bg}.$$
  i = i + 1.
end while
for $1 < i \leq n$ **do**
  $v_i' = v_{i-1}' \circledast m_{i-1}$.
  i = i + 1.
end for

Algorithm 3 Recovery process

Input: De-identified video $V' = \{v_i'\}_{i=1}^n$, key.

Output: Right recovered video $V_r = \{v_{r,i}\}_{i=1}^n$ or wrong recovered video $V_w = \{v_{w,i}\}_{i=1}^n$.

Generate the relative motion flow $\mathcal{M}(V') = \{m_i'\}_{i=1}^{n-1}$

Generate affine transform matrixes T' in Eq. (8.13) and T'^{-1} in Eq. (8.14).

Generate background mask M_{bg}^r with black image M_0:

$$M_{bg}^r = M_0 \cdot T' \cdot T'^{-1}.$$

while $i = 1$ **do**

  $v_1^{p,t} = v_1' \cdot T.$

  if key_is_right **then**

    Recover the right identity embedding r_{rid} with key in Eq. (8.9) and correctly restore the affined first frame:

$$v_{r,1}^{r,t} = \mathcal{F}(r_{rid}, r_{attr}(v_1^{p,t})).$$

    Restore the original layout: $v_{r,1}^r = v_{r,1}^{r,t} \cdot T'^{-1}$.

    Generate right recovered first frame with preserved background:

$$v_{r,1} = v_{r,1}^r \cdot (1 - M_{bg}^r) + v_1' \cdot M_{bg}^r.$$

    i = i + 1.

  else

    Recover a wrong identity embedding r_{wid} with key in Eq. (8.9) and wrongly restore the affined first frame:

$$v_{w,1}^{r,t} = \mathcal{F}(r_{wid}, r_{attr}(v_1^{p,t})).$$

    Restore the original layout: $v_{w,1}^r = v_{w,1}^{r,t} \cdot T'^{-1}$.

    Generate wrong recovered first frame with preserved background:

$$v_{w,1} = v_{w,1}^r \cdot (1 - M_{bg}^r) + v_1' \cdot M_{bg}^r.$$

    i = i + 1.

  end if

end while

for $1 < i \le n$ **do**

  $v_{x,i} = v_{x,i-1} \circledast m_{i-1}', x \in \{r, w\}$.

  i = i + 1.

end for

8.5 Implementation

In this section, we introduce the promising module instantiations and their training process in more detail.

8.5.1 Identity Disentanglement Network Configuration

As mentioned in Sect. 8.4, both the *Protection* and the *Recovery Module* are established on the condition of identity disentanglement. Our identity disentanglement network contains an identity encoder E_{id}, an attribute encoder E_{attr}, and a fusion network $\mathcal{F}$, which are pretrained as a whole on the CelebA-HQ dataset [39].

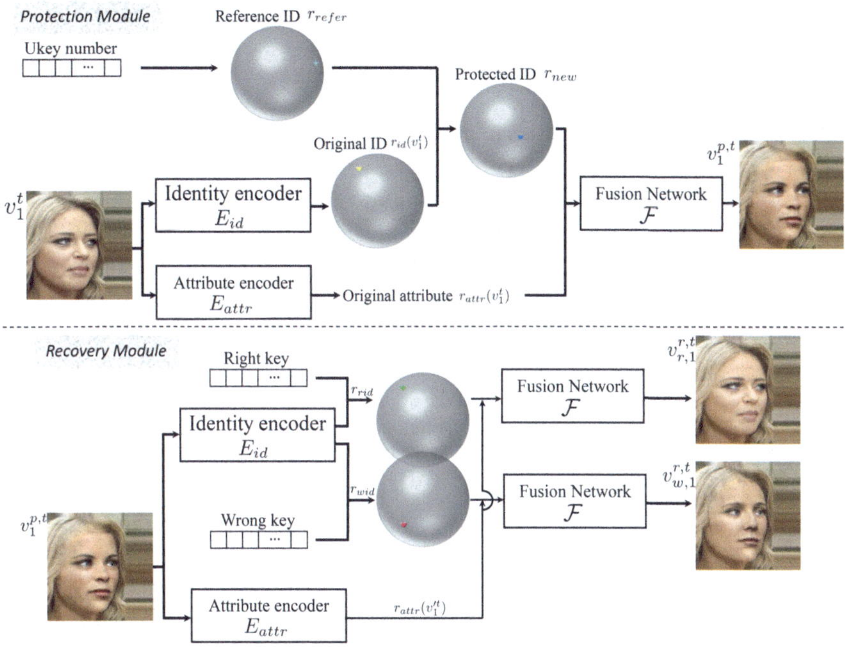

Fig. 8.4 The detailed architecture of the identity disentanglement network in the proposed *Protection Module* and *Recovery Module* with geometrical interpretation of identity changes. Each point on the sphere represents one normalized feature. Different colors denote different identities

Identity Encoder As existing studies on face verification and recognition have made arduous efforts in finding discriminative face features for face identification, we employ a pretrained state-of-the-art (SOTA) face recognition model [40] as our identity encoder. It can provide highly discriminative features for identity verification to avoid training from scratch and has a clear geometric interpretation due to the exact correspondence to the geodesic distance on the hypersphere. Given an original face image X, the identity representation r_{id} is defined to be the last normalized feature vector before the final FC layer, which is denoted as

$$r_{id}(X) = E_{id}(X). \tag{8.22}$$

It is believed that all the embedding features r_{id} are distributed around each feature center on a normalized 512-D hypersphere [40]. Figure 8.4 shows the feature distribution visualization of identity changes. Each point on the sphere represents one normalized feature. Different colors denote different identities.

Attribute Encoder Attribute representation, which determines pose, expression, overall structure, background, and so on, intuitively carries more spatial information than identity. Therefore, in order to preserve different level details, we construct a

Table 8.3 Network structures of identity encoder, attribute encoder, and fusion network

Identity encoder	Attribute encoder			Fusion network
Model [40]			BU ×2	ConvT 4×4,2,1 BN+LR
	Conv 4×4,2,1 BN+LR	CON	ConvT 4×4,2,1 BN+LR	AAD(1024,1024) BU ×2
	Conv 4×4,2,1 BN+LR	CON	ConvT 4×4,2,1 BN+LR	AAD(1024,1024) BU ×2
	Conv 4×4,2,1 BN+LR	CON	ConvT 4×4,2,1 BN+LR	AAD(1024,1024) BU ×2
	Conv 4×4,2,1 BN+LR	CON	ConvT 4×4,2,1 BN+LR	AAD(1024,512) BU ×2
	Conv 4×4,2,1 BN+LR	CON	ConvT 4×4,2,1 BN+LR	AAD(512,256) BU ×2
	Conv 4×4,2,1 BN+LR	CON	ConvT 4×4,2,1 BN+LR	AAD(256,128) BU ×2
	Conv 4×4,2,1 BN+LR	CON	ConvT 4×4,2,1 BN+LR	AAD(128,64) BU ×2
				AAD ResBlk(64,3) BU ×2

Conv 4×4,2,1 represents a convolutional layer with kernel size 4, stride 2, and padding 1. BU represents the BilinearUpsample operation. ConvT 4×4,2,1 represents a transposed convolutional layer with kernel size 4, stride 2, and padding 1. CON represents feature map concatenating. AAD(c_{in}, c_{out}) represents an AAD ResBlk with input and output channels of c_{in} and c_{out}. All LeakyRELUs have $\alpha = 0.1$

U-Net-like structure with a depth of 8, and then use the 8 feature maps generated from the U-Net decoder as the attributes representations r_{attr}. More formally, we denote

$$r_{attr}(X) = E_{attr}(X) = \left\{ r^1_{attr}(X), r^2_{attr}(X), \cdots, r^8_{attr}(X) \right\}, \tag{8.23}$$

where $r^k_{att}(X)$ represents the k-th level feature map from the U-Net decoder.

Fusion Network The fusion network F is required to implement face reconstruction based on r_{id} and r_{attr}. Previous research [41] has verified that direct feature concatenation can easily lead to blurry results and is not expected to be used. To solve this problem, the novel *Adaptive Attentional Denormalization* (AAD) ResBlk [42] has been proposed to improve feature integration in multiple levels. We integrate 8 cascaded AAD ResBlks to the body of our fusion network, in order to adjust the attention regions of r_{id} and r_{attr}, so that they can harmoniously participate in synthesizing different facial parts. And we can get the reconstructed face X' as

$$X' = F(r_{id}(X), r_{attr}(X)). \tag{8.24}$$

The network structure is summarized in Table 8.3.

The whole training process is discussed in the following.

Training Process We use the identity consistency loss $\mathcal{L}_{id}$ to make sure the identity of the reconstructed face $\hat{X}$ still keeps the same:

$$\mathcal{L}_{id} = 1 - \frac{r_{id}(X') \cdot r_{id}(X)}{\|r_{id}(X')\|_2 \cdot \|r_{id}(X)\|_2}. \tag{8.25}$$

Here cosine similarity is chosen because it best fits our angular margin based identity encoder [40].

We also define the attributes consistency loss $\mathcal{L}_{attr}$, which can be formulated as

$$\mathcal{L}_{attr} = \frac{1}{2}\sum_{k=1}^{n} \left\| r_{attr}^{k}(X') - r_{attr}^{k}(X) \right\|_2^2 . \tag{8.26}$$

This loss function has been proved to encourage the generated images to be perceptually similar (but not identical) to the target image [43]. We tried other methods ($\mathcal{L}_1$ distance, Huber loss, and cosine similarity) to measure attributes similarity; however, $\mathcal{L}_2$ distance performs best.

If the restored result X' is generated with the same r_{id} and r_{attr}, it should be as similar to the original image as possible. We set pixel-level $\mathcal{L}_2$ distance as the reconstruction loss:

$$\mathcal{L}_{rec} = \frac{1}{2}\left\| X' - X \right\|_2^2 . \tag{8.27}$$

We take advantage of adversarial learning to train the framework and introduce the adversarial loss $\mathcal{L}_{adv}$ to constrain the generated results indistinguishable from real images. To promote the image quality, it is necessary to expand the perception range of the discriminator, so we adopt m multiscale discriminators [44] with hinge loss functions for different resolution versions of the generated image.

$$\mathcal{L}_{adv}(X_m', X_m) = log(D(X_m)) + log(1 - D(X_m')), \tag{8.28}$$

where X_m indicates the low-resolution image after m-th downsampling.

The total loss function is the weighted sum of the above losses, which can be formulated as

$$\mathcal{L}_{total} = \mathcal{L}_{adv} + \lambda_{attr}\mathcal{L}_{attr} + \lambda_{id}\mathcal{L}_{id} + \lambda_{rec}\mathcal{L}_{rec}, \tag{8.29}$$

where λ_{att}, λ_{id}, and λ_{rec} are the weight parameters for balancing different terms.

In the training process, we use the Adam optimizer [45] with momentum parameters $\beta_1 = 0$, $\beta_2 = 0.999$. The learning rate is set to 4×10^{-4}. The parameters in Eq. (8.29) are set to $\lambda_{att} = \lambda_{rec} = 10$, $\lambda_{id} = 5$.

8.5.2 Other Implementation Details

In the *Motion Flow Module*, we employ a pretrained CNN [46] as our motion flow generator to model the relative dense motion flow. In the *Affine Transformation Module*, we calculate the 5 keypoints (left/right eye, leftmost/rightmost tip of the mouth, and nose) of all faces in CelebA-HQ dataset by [47] and compute the average as the standard point pattern. Then the Umeyama algorithm [48] is utilized to calculate the affine transform matrixes between two point patterns.

8.6 Experiments

8.6.1 Experimental Setup

Dataset We choose the VoxCeleb dataset [49], which contains 22,496 videos extracted from YouTube, to demonstrate the effectiveness of our reversible face video de-identification method. After preprocessing like [46], we obtain 12,775 videos with lengths varying from 64 to 1024 frames, which are resized to 256×256 preserving the aspect ratio. For simplicity, we use the ID number annotated in the dataset as Ukey and define the right key as a number equal to the Ukey, while a random number other than the Ukey is generated as the wrong key.

Comparison Methods To validate the effectiveness of the proposed IdentityMask, we compare to three SOTA methods: ACTION [4], LIVE [5], and CIAGAN [6].

Evaluation Metrics We evaluate the proposed IdentityMask in terms of two metrics, as described below:

(1) Privacy metrics. We measure the cosine similarity of embedding vectors from the generated and original face extracted by pretrained face recognition model, denoted as **CSIM**, to evaluate the quality of identity protection and restoration. For a fair comparison, we employ the well-known FaceNet identification model [50], which is excluded from our training model and pretrained on two public datasets (CASIA-Webface [51] and VGGFace2 [52]), respectively.
(2) Utility metrics. With today's advanced technology, ensuring that the faces in a synthesized video can still be detected is very trivial [53]. Therefore, instead of using the face detection rate, we borrow several metrics that have been commonly used in face swapping and face reenactment tasks. They are designed exactly for videos to evaluate the utility performance. Specifically, the $\mathcal{L}_2$ distances between pose and expression vectors from the generated and original face extracted by an open-sourced pose estimator [54] and a 3D facial model [55] are calculated as pose (denoted as **POSE**) and expression (denoted as **EXP**) similarity. The **FID** score is chosen to evaluate the generation quality as it can measure the distance between the generated distribution and the real distribution. In addition, we evaluate whether the motion of the input video is preserved by computing the average distance of facial landmark keypoints [56] from the generated and original face, which is denoted as **AKD**.

Unless otherwise specified, each metric is calculated independently for each frame.

8.6.2 Comparison in De-identification

In this subsection, we compare our IdentityMask with state-of-the-art face de-identification methods.

The qualitative comparison with ACTION [4] and CIAGAN [6] is shown in Fig. 8.5, while the quantitative results are shown in Table 8.4. It can be seen that the faces generated by ACTION are too visually similar to the original faces, which makes it easy for people to think that they are still the same person and thus does not realize the identity protection from human beings. Besides, incomprehensible artifacts and blurs with light or dark bounding boxes often occur, resulting in obvious video jitter. This makes it difficult to share the generated videos online. In addition, the crucial CSIM value is high, which implies that ACTION is vulnerable to the identification of advanced face verification model.

We can see that the frames generated by CIAGAN can maintain some basic face attributes as well as the rough head orientation, but most of which are not visually similar to the original. These de-identified faces can effectively hide the true identity information from both human eyes and machines. However, distortions

Fig. 8.5 Qualitative comparison on the **VoxCeleb** for face de-identification. The first row shows the original face video frames, and the second to fourth rows show the faces anonymized by ACTION [4], CIAGAN [6], and our method, respectively

Table 8.4 Quantitative comparisons of identity protection on VoxCeleb. The best results are in bold. ↑ means higher is better, and ↓ means lower is better

Method	CSIM↓		POSE↓	EXP↓	FID↓	AKD↓
	CASIA	VGGFace2				
ACTION	0.904	0.869	**2.45**	2.69	19.34	1.60
CIAGAN	0.520	0.507	14.69	8.23	31.50	4.16
Ours	**0.518**	**0.503**	**2.45**	**2.64**	**18.70**	**1.58**

and artifacts often occur. When the characters perform large poses or expressions, there will even be large deformation. These are very unfavorable to the video. As can be seen from Table 8.4, its utility metrics deteriorate significantly, which will make the synthesized videos hard to meet the requirement for online sharing.

In contrast, our method produces more natural looking images that achieve a great advantage in visual similarity to the input frame with visually perceptible changes and enables de-identification for both human beings and machines. Furthermore, from Table 8.4, the lowest CSIM value indicates that our method is superior to the compared method in protecting the real identity. Meanwhile, the best performance under utility metrics shows that our method also well preserves the nonidentity aspects of the original frame, i.e., pose, expression, facial motion, and overall structure. So we can best ensure the subsequent normal use of the de-identified faces.

A comparison with the work of LIVE [5] is given in Fig. 8.6. Our results are at least visually as good as the original ones, despite having to run on the cropped faces extracted from the paper PDF.

To make the comparison more convincing and fairer, we follow the evaluation protocol that has been used in [5] and [6], which is conducted on the LFW benchmark. Specifically, two FaceNet identification models (pretrained on CASIA-Webface and VGGFace2, respectively) are employed, and the main evaluation metric is the true acceptance rate. Table 8.5 presents the results on de-identified LFW image pairs for a given person, while the de-identification method is applied to the second image of each pair. It can be seen that all methods can significantly reduce the true positive rate. In particular, our method achieves the best privacy protection.

Fig. 8.6 Comparison of the de-identified faces between LIVE [5] and our method

Table 8.5 Quantitative evaluation with Sota methods on LFW datasets

Method	True positive rate↓ CASIA	VGGFace2
Original	0.965 ± 0.016	0.986 ± 0.010
ACTION	0.696 ± 0.015	0.714 ± 0.014
LIVE	0.035 ± 0.011	0.038 ± 0.015
CIAGAN	0.019 ± 0.008	0.034 ± 0.015
Ours	**0.017 ± 0.011**	**0.026 ± 0.014**

8.6.3 Analysis in Identity Recovery

In this subsection, we evaluate our performance in identity restoration. The effect of one original video being de-identified and recovered, respectively, with the right and wrong keys (denoted as "R_key" and "W_key") is presented in Fig. 8.7. It can be seen that the identity-protected frames obtain a new identity, while still maintaining a high visual similarity (i.e., appearance, pose, expression, and facial motion), which ensures the rationality of subsequent use. Then the right key can restore a video that is exactly similar to the original video with the real identity, while the wrong key can restore a realistic video with another new identity different from the original identity. Moreover, each wrong key maps to a unique identity. In this way, we provide security via ambiguity: Even if a privacy intruder guesses the correct key, it is extremely difficult to know that without having access to any other identity revealing meta-data, since each key—regardless of whether it is correct or not—always leads to a different realistic identity. In particular, the effect of being recovered by multiple wrong keys is shown in Fig. 8.8.

The quantitative results of identity recovery are shown in Table 8.6. It shows that after de-identification: (1) The original identity can be recovered excellently with

Fig. 8.7 Qualitative results of our method about identity protection and identity recovery on the VoxCeleb dataset

Fig. 8.8 Qualitative results of identity recovery when given multiple wrong keys. The black background indicates the original videos, and the red background indicates the wrong recovered videos

Table 8.6 Quantitative results of identity recovery on VoxCeleb

	CSIM					
Method	CASIA	VGGFace2	POSE↓	EXP↓	FID↓	AKD↓
R_key	0.961	0.959	1.62	1.59	8.50	1.34
W_key	0.475	0.461	2.75	2.96	23.18	1.61

Table 8.7 Quantitative experimental results of right recovery quality on VoxCeleb

	LPIPS↓	PSNR↑	SSIM↑	MAE↓
R_key	0.077	25.492	0.875	0.036

the correct key, which is conducive to the supervision of network abnormal events; (2) When given the wrong key, it is almost impossible to restore the original identity; (3) Whether the video is recovered by the right or wrong key, its utility is always impressive.

To better evaluate the right recovery quality, we apply LPIPS (learned perceptual image patch similarity) distance [57] to measure perceptual similarity, PSNR (peak signal-to-noise ratio) [58], and MAE (mean absolute error) to measure distortion at the pixel level, and SSIM (structural similarity) [59] to measure the structure similarity. The results in Table 8.7 demonstrate that the right recovered frames are extremely similar to the original frames, which is consistent with the intuitive expectation. To the best of our knowledge, IdentityMask is the first work to achieve de-identified face video restoration, so the above results are summarized as the baseline for future research.

8.6.4 Model Analysis and Discussions

In this subsection, considering the de-identification process $V \rightarrow V'$ and the recovery process $V' \rightarrow, V$ are symmetrical, while previous comparison methods can only do de-identification, we take the former as the example:

(1) *Motion Flow Module Selection.* We pick and compare three SOTA motion flow modelling methods: *STD* [46], *RLT* [46], and *AVD* [60]. *STD* computes the deep motion flow between input and output videos frame by frame. *RLT* calculates the deep motion flow between every two adjacent frames of the input video first and then applies this relative dense motion flow to the first frame of the output video. *AVD* also computes the deep motion flow between input and output videos frame by frame, except it disentangles the shape and pose of objects in the region space and forces decoupling of foreground from background. Different motion flow modelling methods are suitable for different application scenarios. *STD* directly transfers object shape from the input video into the generated video, while the *RLT* requires that objects be in the same pose in the first frame of the input and output videos, and the *AVD* is designed specifically

Fig. 8.9 Quantitative comparison on the **Voxceleb** of the influence of different *Motion Flow Module*. The first row is the original frames. The second to fourth rows demonstrate results when using *STD* [46], *AVD* [60], and *RLT* [46]

Table 8.8 Quantitative evaluation of the Motion Flow Module

Method	CSIM↓		POSE↓	EXP↓	FID↓	AKD↓
	CASIA	VGGFace2				
STD	0.522	0.513	**2.45**	**2.62**	20.96	1.59
AVD	0.520	0.507	2.53	2.78	22.11	1.61
RLT(ours)	**0.518**	**0.503**	**2.45**	2.64	**18.70**	**1.58**

for videos of articulated objects. Since the face is exactly in the same pose in the first frame of the input and synthesized video in either de-identification process or recovery process, the *RLT* is theoretically the best motion flow modelling method for IdentityMask. Figure 8.9 and Table 8.8 reveal the qualitative and quantitative results of the influence of the *Motion Flow Module*. We can see that if *STD* is used, the synthesized face can retain a slightly more similar expression to the original face than using *RLT*. However, the transfer of the original face shape leads to a decline of privacy protection ability, while lower FID and AKD values also indicate poorer generation quality and motion flow modelling. In addition, the background of the generated face has severe distortion. If *AVD* is used, not only the eyes and mouth have obvious distortion, but also all the quantitative metrics are worse than employing *RLT*. Therefore, *RLT* is the best choice of our *Motion Flow Module*.

(2) *Ablation Study*. We take two variants of the proposed IdentityMask pipeline for ablation study in order to validate effectiveness of *Affine Transformation Module* and *Motion Flow Module*. Specifically, *w/o AT* indicates the variant without the *Affine Transformation Module*, and *w/o MF* indicates the variant without the *Motion Flow Module*, which means that the input videos have to be processed frame by frame. We let "ours" indicate the full model. Figure 8.10 shows the qualitative results and Table 8.9 presents the quantitative comparison. It can be seen that *w/o AT* generates a very casual face contour and results in a substantial decline in data utility. This is unacceptable for media users. As for *w/o MF*, although it can protect identity slightly better than the full

Fig. 8.10 Ablation study on the **Voxceleb** of our method. The first row is the original frames, the second row to the fourth row show the corresponding de-identified results of *w/o AT* (the model without the *Affine Transformation Module*), *w/o MF* (the model without the *Motion Flow Module*) and the full model

Table 8.9 Ablation study of the proposed IdentityMask pipeline

Method	CSIM↓ CASIA	VGGFace2	POSE↓	EXP↓	FID↓	AKD↓
w/o AT	**0.376**	**0.368**	22.58	10.27	43.04	3.96
w/o MF	0.513	0.492	2.50	2.71	20.47	2.24
Ours	0.518	0.503	**2.45**	**2.64**	**18.70**	**1.58**

model, its pose and expression are less similar to the original video. Also, its image quality is poor, especially the preservation of facial movements. These will render the synthesized video unfavorable for subsequent identity-agnostic use. Therefore, each module in our method is indispensable, and only the full model can achieve the most wonderful de-identified effects without affecting the subsequent identity-agnostic use.

(3) *Parameter Selection.* In this subsection, the performance variation of de-identification with respect to the controllable parameter θ is studied. We conduct a group of identity protection experiments with respect to the parameter θ. During the test, we randomly select 1000 videos from the VoxCeleb dataset and change θ from 0 to 90 for each video to synthesize the corresponding de-identified videos. Figure 8.11 shows the qualitative results. It can be observed that with the increase of θ, the visual identity difference between the synthetic faces and the original faces expands, while the identity-independent attributes are still maintained. Here, both the privacy metrics and the utility metrics are used to evaluate the overall identity protection effect and are shown in Fig. 8.12. It can be seen that the degree of identity protection can be adjusted, accompanied by utility variations. Considering the identity protection effect and the utility performance comprehensively, we set θ to 60 for all other experiments.

(4) *Computational Overhead Analysis.* We explore the contribution of the *Motion Flow Module* to saving computational overheads in de-identification tasks. We

Fig. 8.11 De-identified results with variant parameter θ values

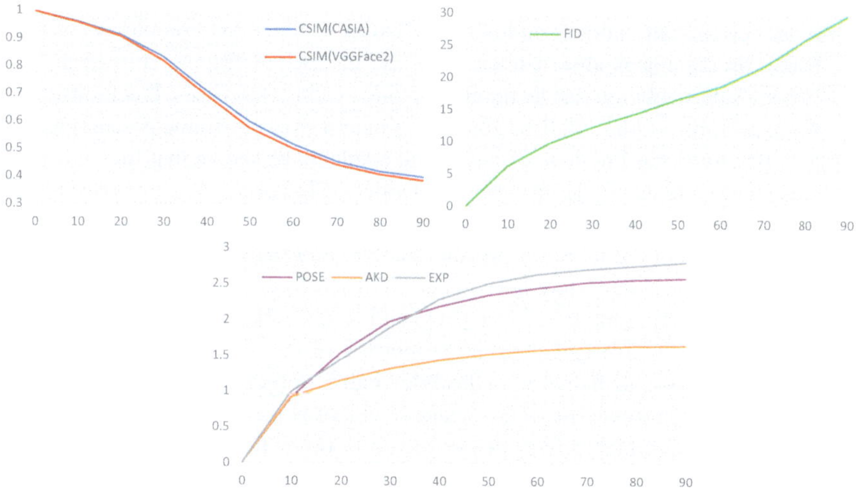

Fig. 8.12 The performance variation of de-identification with respect to the parameter θ. The x-axis indicates θ value, and the y axis indicates the metric values

compare with ACTION and CIAGAN on an NVIDIA GTX 1080 Ti, and the results are shown in Fig. 8.13. We observe that when the number of video frames is greater than 80, our method has a lower computational complexity than ACTION, and when the number of video frames is greater than 360, our method

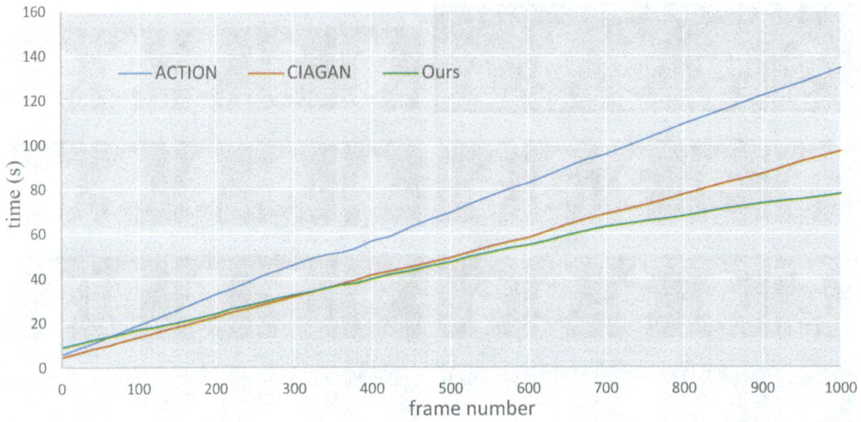

Fig. 8.13 Comparison of computational overheads between ACTION [4], CIAGAN [6], and our method on the VoxCeleb dataset

Table 8.10 Security analysis of the proposed IdentityMask

Method	CViT [61]	LRNet [62]
Fake video detection rate	74.8%	63.1%

is less computationally complex than CIAGAN. Since the complexity of per-frame processing is almost linear with the number of frames, this advantage becomes more obvious as the number of video frames increases. It demonstrates the superiority of motion flow guided evaluation over per-frame processing.

(5) *Security Analysis*. Previous experimental results have shown that IdentityMask can generate realistic identity-protected videos. However, we are worried about the potential misuse. Once abused, even if the authority can obtain the real identity through the recovery process, unpleasant effects (such as fraud) in the dissemination process may have occurred. Therefore, we apply two advanced deepfake detection models, CViT [61] and LRNet [62], to examine the security of IdentityMask. We calculate the proportion of de-identified videos that are judged as fake and name it as the **fake video detection rate**. As shown in Table 8.10, the probability of the synthetic videos being judged as "deepfake" is relatively high, which proves that our identity protection technology has good security despite its SOTA utility performance.

8.7 Conclusions

In this chapter, we have proposed a reversible face video de-identification framework, IdentityMask, guided by deep motion flow. Our framework consists of a de-identification process and a recovery process. The former is able to conceal the real identity with a visually similar appearance in a seamless way, and the latter aims to recover the original identity only when given the right key. The proposed framework is the first one suitable for reversible face video de-identification. It presents a quality that surpasses the literature methods in the de-identification task and is impressive in the identity recovery process. Besides, instead of existing per-frame processing, we take advantage of motion flow to guide consecutive frames generation, which alleviates the computational overhead and improves the synthesis effect. Extensive experimental results on a standard diverse dataset verify the effectiveness and efficiency of our framework.

While our reversible face video de-identification results are visibly convincing, additional improvements are possible. As part of our future work, we plan to elaborate the mapping function between Ukey and right key to further enhance the security of identity protection.

References

1. J. Zhou, C.-M. Pun, Y. Tong, Privacy-sensitive objects pixelation for live video streaming, in *Proceedings of the 28th ACM International Conference on Multimedia* (2020), pp. 3025–3033
2. S. Jia, X. Li, C. Hu, G. Guo, Z. Xu, 3d face anti-spoofing with factorized bilinear coding. IEEE Trans. Circuits Syst. Video Technol. **31**(10), 4031–4045
3. B. Liu, M. Ding, S. Shaham, W. Rahayu, F. Farokhi, Z. Lin, When machine learning meets privacy: a survey and outlook. ACM Comput. Surv. **54**(2), 1–36 (2021)
4. Z. Ren, Y. J. Lee, M.S. Ryoo, Learning to anonymize faces for privacy preserving action detection, in *Proceedings of the European Conference on Computer Vision (ECCV)* (2018), pp. 620–636
5. O. Gafni, L. Wolf, Y. Taigman, Live face de-identification in video, in *Proceedings of the IEEE/CVF International Conference on Computer Vision* (2019), pp. 9378–9387
6. M. Maximov, I. Elezi, L. Leal-Taixé, Ciagan: conditional identity anonymization generative adversarial networks, in *Proceedings of the IEEE/CVF Conference on Computer Vision and Pattern Recognition* (2020), pp. 5447–5456
7. H. Wu, X. Tian, M. Li, Y. Liu, G. Ananthanarayanan, F. Xu, S. Zhong, Pecam: privacy-enhanced video streaming and analytics via securely-reversible transformation, in *Proceedings of the 27th Annual International Conference on Mobile Computing and Networking* (2021), pp. 229–241
8. R. Gross, L. Sweeney, J. Cohn, F. De la Torre, S. Baker, Face de-identification, in *Protecting Privacy in Video Surveillance* (Springer, Berlin, 2009), pp. 129–146
9. B. Samarzija, S. Ribaric, An approach to the de-identification of faces in different poses, in *2014 37th International Convention on Information and Communication Technology, Electronics and Microelectronics (MIPRO)* (IEEE, 2014), pp. 1246–1251
10. B. Meden, R.C. Mallı, S. Fabijan, H.K. Ekenel, V. Štruc, P. Peer, Face deidentification with generative deep neural networks. IET Signal Process. **11**(9), 1046–1054 (2017)
11. Y. Li, S. Lyu, De-identification without losing faces, in *Proceedings of the ACM Workshop on Information Hiding and Multimedia Security* (2019), pp. 83–88
12. B. Zhu, H. Fang, Y. Sui, L. Li, Deepfakes for medical video de-identification: privacy protection and diagnostic information preservation, in *Proceedings of the AAAI/ACM Conference on AI, Ethics, and Society* (2020), pp. 414–420
13. Y. Wen, B. Liu, R. Xie, J. Cao, L. Song, Deep motion flow aided face video de-identification, in *2021 IEEE International Conference on Visual Communications and Image Processing (VCIP)*

14. J. Cao, B. Liu, Y. Wen, R. Xie, L. Song, Personalized and invertible face de-identification by disentangled identity information manipulation, in *Proceedings of the IEEE/CVF International Conference on Computer Vision* (2021), pp. 3334–3342
15. W. Sun, J. Zhou, Y. Li, M. Cheung, J. She, Robust high-capacity watermarking over online social network shared images. IEEE Trans. Circuits Syst. Video Technol. **31**(3), 1208–1221 (2020)
16. L. Wu, Y. Wang, H. Yin, M. Wang, L. Shao, Few-shot deep adversarial learning for video-based person re-identification. IEEE Trans. Image Process. **29**, 1233–1245 (2019)
17. L. Wu, R. Hong, Y. Wang, M. Wang, Cross-entropy adversarial view adaptation for person re-identification. IEEE Trans. Circuits Syst. Video Technol. **30**(7), 2081–2092 (2019)
18. F. Dufaux, T. Ebrahimi, Scrambling for privacy protection in video surveillance systems. IEEE Trans. Circuits Syst. Video Technol. **18**(8), 1168–1174 (2008)
19. J. Schiff, M. Meingast, D.K. Mulligan, S. Sastry, K. Goldberg, Respectful cameras: detecting visual markers in real-time to address privacy concerns, in *Protecting Privacy in Video Surveillance* (Springer, Berlin, 2009), pp. 65–89
20. D. Chen, Y. Chang, R. Yan, J. Yang, Protecting personal identification in video, in *Protecting Privacy in Video Surveillance* (Springer, Berlin, 2009), pp. 115–128
21. P. Agrawal, P. Narayanan, Person de-identification in videos. IEEE Trans. Circuits Syst. Video Technol. **21**(3), 299–310 (2011)
22. M. Mrityunjay, P. Narayanan, The de-identification camera, in *2011 Third National Conference on Computer Vision, Pattern Recognition, Image Processing and Graphics (NCVPRIPG)* (2011), pp. 192–195
23. M. Ivasic-Kos, A. Iosifidis, A. Tefas, I. Pitas, Person de-identification in activity videos, in *2014 37th International Convention on Information and Communication Technology, Electronics and Microelectronics (MIPRO)* (IEEE, Berlin, 2014), pp. 1294–1299
24. M. Blažević, K. Brkić, T. Hrkać, Towards reversible de-identification in video sequences using 3d avatars and steganography (2015). arXiv preprint arXiv:1510.04861
25. M.S. Ryoo, B. Rothrock, C. Fleming, H.J. Yang, Privacy-preserving human activity recognition from extreme low resolution, in *Thirty-First AAAI Conference on Artificial Intelligence* (2017)
26. E. Flouty, O. Zisimopoulos, D. Stoyanov, Faceoff: anonymizing videos in the operating rooms, in *OR 2.0 Context-Aware Operating Theaters, Computer Assisted Robotic Endoscopy, Clinical Image-Based Procedures, and Skin Image Analysis* (Springer, Berlin, 2018), pp. 30–38
27. M. Yamaç, M. Ahishali, N. Passalis, J. Raitoharju, B. Sankur, M. Gabbouj, Reversible privacy preservation using multi-level encryption and compressive sensing, in *2019 27th European Signal Processing Conference (EUSIPCO)* (IEEE, 2019), pp. 1–5
28. M.U. Kim, H. Lee, H.J. Yang, M.S. Ryoo, Privacy-preserving robot vision with anonymized faces by extreme low resolution, in *2019 IEEE/RSJ International Conference on Intelligent Robots and Systems (IROS)* (IEEE, 2019), pp. 462–467
29. Z.W. Wang, V. Vineet, F. Pittaluga, S.N. Sinha, O. Cossairt, S. Bing Kang, Privacy-preserving action recognition using coded aperture videos, in *Proceedings of the IEEE/CVF Conference on Computer Vision and Pattern Recognition Workshops* (2019)
30. N.A. Tu, K.-S. Wong, M.F. Demirci, Y.-K. Lee et al., Toward efficient and intelligent video analytics with visual privacy protection for large-scale surveillance. J. Supercomput. **77**(12), 14374–14404
31. T. Winkler, B. Rinner, Trustcam: security and privacy-protection for an embedded smart camera based on trusted computing, in *2010 7th IEEE International Conference on Advanced Video and Signal Based Surveillance* (IEEE, Berlin, 2010), pp. 593–600
32. A. Erdélyi, T. Barát, P. Valet, T. Winkler, B. Rinner, Adaptive cartooning for privacy protection in camera networks, in *2014 11th IEEE International Conference on Advanced Video and Signal Based Surveillance (AVSS)* (IEEE, 2014), pp. 44–49
33. K. Brkić, T. Hrkać, Z. Kalafatić, Protecting the privacy of humans in video sequences using a computer vision-based de-identification pipeline. Expert Syst. Appl. **87**, 41–55 (2017)
34. S. Çiftçi, A. O. Akyüz, T. Ebrahimi, A reliable and reversible image privacy protection based on false colors. IEEE Trans. Multimedia **20**(1), 68–81 (2017)

35. X. Ben, C. Gong, P. Zhang, R. Yan, Q. Wu, W. Meng, Coupled bilinear discriminant projection for cross-view gait recognition. IEEE Trans. Circuits Syst. Video Technol. **30**(3), 734–747 (2019)
36. J. Li, L. Han, H. Zhang, X. Han, J. Ge, X. Cao, Learning disentangled representations for identity preserving surveillance face camouflage, in *2020 25th International Conference on Pattern Recognition (ICPR)* (IEEE, 2021), pp. 9748–9755
37. H. Proença, The uu-net: reversible face de-identification for visual surveillance video footage. IEEE Trans. Circuits Syst. Video Technol. **32**(2), 496–509 (2021)
38. X. Zhu, Y. Xiong, J. Dai, L. Yuan, Y. Wei, Deep feature flow for video recognition, in *Proceedings of the IEEE Conference on Computer Vision and Pattern Recognition* (2017), pp. 2349–2358
39. T. Karras, T. Aila, S. Laine, J. Lehtinen, Progressive growing of gans for improved quality, stability, and variation (2017). arXiv preprint arXiv:1710.10196
40. J. Deng, J. Guo, N. Xue, S. Zafeiriou, Arcface: additive angular margin loss for deep face recognition, in *Proceedings of the IEEE Conference on Computer Vision and Pattern Recognition* (2019), pp. 4690–4699
41. J. Bao, D. Chen, F. Wen, H. Li, G. Hua, Towards open-set identity preserving face synthesis, in *Proceedings of the IEEE Conference on Computer Vision and Pattern Recognition* (2018), pp. 6713–6722
42. L. Li, J. Bao, H. Yang, D. Chen, F. Wen, Faceshifter: towards high fidelity and occlusion aware face swapping (2019). arXiv preprint arXiv:1912.13457
43. J. Johnson, A. Alahi, L. Fei-Fei, Perceptual losses for real-time style transfer and super-resolution, in *European Conference on Computer Vision* (Springer, Berlin, 2016), pp. 694–711
44. T. Park, M.-Y. Liu, T.-C. Wang, J.-Y. Zhu, Semantic image synthesis with spatially-adaptive normalization, in *Proceedings of the IEEE Conference on Computer Vision and Pattern Recognition* (2019), pp. 2337–2346
45. D.P. Kingma, J. Ba, Adam: a method for stochastic optimization (2014). arXiv preprint arXiv:1412.6980
46. A. Siarohin, S. Lathuilière, S. Tulyakov, E. Ricci, N. Sebe, First order motion model for image animation. Adv. Neural Inform. Process. Syst. **32**, 7137–7147 (2019)
47. K. Zhang, Z. Zhang, Z. Li, Y. Qiao, Joint face detection and alignment using multitask cascaded convolutional networks. IEEE Signal Process. Lett. **23**(10), 1499–1503 (2016)
48. S. Umeyama, Least-squares estimation of transformation parameters between two point patterns. IEEE Comput. Archit. Lett. **13**(4), 376–380 (1991)
49. A. Nagrani, J.S. Chung, A. Zisserman, Voxceleb: a large-scale speaker identification dataset (2017). arXiv preprint arXiv:1706.08612
50. F. Schroff, D. Kalenichenko, J. Philbin, Facenet: a unified embedding for face recognition and clustering, in *Proceedings of the IEEE Conference on Computer Vision and Pattern Recognition* (2015), pp. 815–823
51. D. Yi, Z. Lei, S. Liao, S. Z. Li, Learning face representation from scratch (2014). arXiv preprint arXiv:1411.7923
52. Q. Cao, L. Shen, W. Xie, O.M. Parkhi, A. Zisserman, Vggface2: a dataset for recognising faces across pose and age, in *2018 13th IEEE International Conference on Automatic Face & Gesture Recognition (FG 2018)* (IEEE, 2018), pp. 67–74
53. H. Wu, G. Liu, Y. Yao, X. Zhang, Watermarking neural networks with watermarked images. IEEE Trans. Circuits Syst. Video Technol. **31**(7), 2591–2601
54. N. Ruiz, E. Chong, J.M. Rehg, Fine-grained head pose estimation without keypoints, in *Proceedings of the IEEE Conference on Computer Vision and Pattern Recognition Workshops* (2018), pp. 2074–2083
55. Y. Deng, J. Yang, S. Xu, D. Chen, Y. Jia, X. Tong, Accurate 3d face reconstruction with weakly-supervised learning: from single image to image set, in *Proceedings of the IEEE/CVF Conference on Computer Vision and Pattern Recognition Workshops* (2019)

56. A. Bulat, G. Tzimiropoulos, How far are we from solving the 2d & 3d face alignment problem? (and a dataset of 230,000 3d facial landmarks), in *Proceedings of the IEEE International Conference on Computer Vision* (2017), pp. 1021–1030
57. R. Zhang, P. Isola, A.A. Efros, E. Shechtman, O. Wang, The unreasonable effectiveness of deep features as a perceptual metric, in *Proceedings of the IEEE Conference on Computer Vision and Pattern Recognition* (2018), pp. 586–595
58. Q. Huynh-Thu, M. Ghanbari, Scope of validity of psnr in image/video quality assessment. Electron. Lett. **44**(13), 800–801 (2008)
59. Z. Wang, A.C. Bovik, H.R. Sheikh, E.P. Simoncelli, Image quality assessment: from error visibility to structural similarity. IEEE Trans. Image Process. **13**(4), 600–612 (2004)
60. A. Siarohin, O.J. Woodford, J. Ren, M. Chai, S. Tulyakov, Motion representations for articulated animation, in *Proceedings of the IEEE/CVF Conference on Computer Vision and Pattern Recognition* (2021), pp. 13653–13662
61. D. Wodajo, S. Atnafu, Deepfake video detection using convolutional vision transformer (2021). arXiv preprint arXiv:2102.11126
62. Z. Sun, Y. Han, Z. Hua, N. Ruan, W. Jia, Improving the efficiency and robustness of deepfakes detection through precise geometric features, in *Proceedings of the IEEE/CVF Conference on Computer Vision and Pattern Recognition* (2021), pp. 3609–3618

Part III
Conclusion and Future Work

Chapter 9
Future Prospects and Challenges

9.1 Future Prospects and Open Research Questions

In this section, we provide a summary of the main challenges and open issues associated with face de-identification methods and outline potential future directions.

Privacy Guarantees with High Utility Ensuring provable privacy is a crucial aspect of face de-identification methods [1]. Existing privacy metrics like k-anonymity are based on strong assumptions and may not adequately address real-world problems [2]. Combining privacy theory with deep learning technology to enhance the measurement and interpretability of de-identification algorithms is an important avenue for improvement [3]. In addition, striking the right balance between preserving the utility of the data and ensuring privacy is a constant challenge. De-identification techniques should provide sufficient protection while retaining essential information for legitimate use cases.

Evaluation Metrics Despite the summary of commonly used evaluation metrics in Chap. 3, existing metrics often fail to comprehensively quantify the performance of de-identification algorithms. Many utility metrics rely on the similarity between de-identified results and the original images. The absence of a universally accepted evaluation system or indicator for de-identification highlights the need for proposing new evaluation criteria tailored to the characteristics of these algorithms, which will drive further development in this field.

Generalization Capability of Algorithms Some algorithms rely on the pretrained face recognition model to obtain identity embeddings, especially *Generative Model-Based Identity Modification* algorithms [4–8] and adversarial perturbation [9]. While the former group demonstrates better generalization, the latter is often effective only for specific face recognition models. Moreover, deep neural networks (DNNs) are heavily influenced by datasets. Face de-identification models should

Y. Wen et al., *Face De-identification: Safeguarding Identities in the Digital Era*,
https://doi.org/10.1007/978-3-031-58222-6_9

generalize well across different datasets. Ensuring that a model trained on one dataset can effectively de-identify faces in another dataset is a significant challenge.

Integration with Real-Life Applications The de-identification algorithms mainly target the image processing stage, while privacy protection can be achieved from image acquisition, storage, publishing, and other aspects through a more complete system design in reality. It is also very meaningful for the development of related deep learning technologies to generate datasets without privacy threats by face de-identification technology. On the other hand, applying face de-identification techniques in real-world, dynamic settings, such as video streams and live surveillance [10, 11], presents unique challenges. Ensuring real-time de-identification while maintaining accuracy is a complex task.

Controllable and Fine-Grained Privacy Different levels of de-identification are expected in distinctive scenarios, and the needs of various users are disparate even in the same scene. Therefore, an ideal face de-identification method should be adjustable and controllable to adapt to a wide range of application scenarios. Separating the protection process from the network training may be an effective way to improve the flexibility of the algorithm. Considering special circumstances such as crime tracking, we prefer to use the original images rather than the de-identified ones in some cases. At present, most studies focus on the protection process, while only a few can achieve recoverable de-identification. We think it is also meaningful to consider the restoration process in the future.

Moreover, achieving more fine-grained processing focused on identity protection can lead to enhanced effectiveness and utility. Separation and operation of identity-related elements can offer better protection under similar levels of disturbance. The challenge here lies in disentangling identity from attributes, an active area of research that includes exploring more explicit latent spaces, developing comprehensive identity representations, and introducing contrast loss in training.

Ethical and Regulatory Compliance Addressing the ethical implications of face de-identification is vital. Researchers must consider the broader societal impacts of their work, such as potential biases and fairness issues. In addition, keeping up with evolving privacy regulations and ensuring that de-identification methods comply with these regulations can be challenging. Researchers need to stay informed about legal requirements.

Broader De-identification on Other Biometric Information With the development of biometric recognition technology, the information that can be used for identification is not limited to face images [12]. Beyond faces, extending de-identification techniques to other biometric modalities also poses interesting challenges:

- Voice de-identification—Removing identifying vocal cues while preserving naturalness and intelligibility of speech is difficult.
- Gait de-identification [13]—Anonymizing motion patterns in video while maintaining natural walking/running styles is an open problem.

- Fingerprint de-identification—Obscuring unique patterns in fingerprints or iris scans in a reversible way for authentication is unsolved.
- Multimodal de-identification—Jointly de-identifying faces, voice, gait, etc., in a coordinated way to maximize anonymity remains largely unexplored.
- Cross-modal de-identification—Transferring de-identification across modalities, e.g., de-identified faces to voices, presents cross-domain challenges.

9.2 Technical Challenges

9.2.1 Low-Complexity and Real-Time De-identification Methods

Improving the efficiency and real-time performance of face de-identification models would enable many more valuable applications. This might be achieved by holistically combining model, software and hardware optimizations tailored for real-time usage. But there are technical challenges to be solved:

- Model Compression Tradeoffs—Techniques like pruning and quantization can improve efficiency but may hurt model accuracy. Balancing compression rate, speedup, and maintained accuracy is challenging.
- Hardware Constraints—Optimizing for target hardware like smartphones and embedded devices with limited memory, compute, and power is difficult. Models need hardware-aware codesign.
- Generalizability vs. Efficiency—Highly compressed models may fail to generalize well to new data distributions. Maintaining robustness with efficiency is tricky.
- Software Optimizations—Model efficiency alone is not sufficient. Efficient pre/postprocessing and software optimization are crucial but difficult.

Overall, balancing speed, accuracy, robustness, and fairness simultaneously for real-time face de-identification remains an impactful yet challenging research direction.

9.2.2 Preventing Reverse Engineering Attacks of De-identified Faces

Protecting de-identified faces from adversarial attacks that aim to reverse engineer or reconstruct the original identity. Here are some key technical challenges:

- Generative Model-Based Attacks—Generative models like GANs and diffusion models can generate highly realistic faces. Defending against generative adversarial network (GAN)-powered reconstruction attacks is an arms race.

- Cross-Database Recognition—Even without exact reconstruction, linking de-identified faces to identities in other databases poses risks. Avoiding cross-DB leaks is tricky.
- Reidentification—Using auxiliary information, attackers could reidentify anonymized faces. Real-world environments and data distributions make reversal attacks more feasible. Developing robust and provable anonymity is an open problem.

In summary, developing theoretically grounded, empirically robust defenses against adversarial reversed engineering of face de-identification systems in real-world conditions remains a significant open research problem.

9.2.3 Moving Beyond Supervised Learning on Limited Datasets

Overreliance on supervised learning is a limitation for advancing face de-identification research. Here are some key technical challenges in moving beyond supervised learning:

- Limited Labelled Data—Collecting large datasets with labels for de-identification is expensive and impedes progress. Reducing reliance on labelled data is key.
- Data Imbalance—Performance of de-identification methods can vary widely across different demographic groups, image types, etc. Generalizable techniques that work across diverse real-world data are needed. Biased datasets cause poor minority group performance. Correcting imbalance without labels is an open problem.
- Semi/self-supervised learning: Consistency regularization strategies for semisupervised learning have hyperparameters that require tuning, while for self-supervised learning, pretraining tasks like contrastive learning may still require large labelled datasets for fine-tuning.

In summary, safely and effectively leveraging unlabelled or weakly labelled data for model generalization, robustness, and fairness—while still quantifying performance—remains a key challenge for advancing face de-identification.

9.2.4 Multimodal De-identification

Multimodal de-identification, involving anonymizing multiple biometrics like face, voice, gait, etc., together, poses some unique technical challenges:

- Correlated Modalities and Cross-Modal Consistency—Face, voice, and body language are often highly correlated. Jointly decorrelating them is tricky. On the other hand, maintaining natural coherence across de-identified modalities is challenging.

- Heterogeneous Data and Domain Gap—Synchronizing different frame rates, resolutions, and capture conditions for multibiometric data is difficult. Also, adaptively understanding and preserving the minimum identifying information across modalities is an open problem. In addition, joint distributions seen during training may differ significantly from deployment, causing coherence failures.
- Disentangled Representations—Learning disentangled representations for isolating and selectively de-identifying identity factors poses challenges.
- Computational Complexity—Multistream models for multiple biometrics can be resource-intensive.
- Security—With multiple points of attack, ensuring equivalent protections across modalities is nontrivial.

Robustly de-identifying multiple biometrics while preserving utility and naturalness in real-world conditions remains an open multifaceted research problem requiring interdisciplinary perspectives.

References

1. Y. Wen, B. Liu, M. Ding, R. Xie, I. L. Song, IdentityDP: differential private identification protection for face images. Neurocomputing **501**, 197–211 (2022)
2. J. Cao, B. Liu, Y. Wen, Y. Zhu, R. Xie, L. Song, Hiding among your neighbors: face image privacy protection with differential private k-anonymity, in *Proceedings of the IEEE International Symposium on Broadband Multimedia Systems and Broadcasting (BMSB)* (IEEE, Piscataway, 2022), pp. 1–6
3. Y. Wen, B. Liu, J. Cao, R. Xie, L. Song, Divide and conquer: A two-step method for high quality face de-identification with model explainability, in *Proceedings of the IEEE/CVF International Conference on Computer Vision (ICCV)* (2023), pp. 269–272
4. J. Cao, B. Liu, Y. Wen, R. Xie, L. Song, Personalized and invertible face de-identification by disentangled identity information manipulation, in *ICCV* (2021)
5. Y. Zhao, B. Liu, T. Zhu, M. Ding, W. Zhou, Private-encoder: enforcing privacy in latent space for human face images. Concurr. Comput. Practice Experience **34**(3), e6548 (2021)
6. B. Liu, M. Ding, H. Xue, T. Zhu, D. Ye, L. Song, W. Zhou, DP-image: differential privacy for image data in feature space (2021). arXiv preprint arXiv:2103.07073
7. Y. Wen, B. Liu, R. Xie, Y. Zhu, J. Cao, L. Song, A hybrid model for natural face de-identification with adjustable privacy, in *2020 IEEE International Conference on Visual Communications and Image Processing, VCIP 2020* pp. 269–272
8. J. Yu, H. Xue, B. Liu, Y. Wang, S. Zhu, M. Ding, GAN-based differential private image privacy protection framework for the internet of multimedia things. Sensors 21(1), 58 (2021)
9. H. Xue, B. Liu, X. Yuan M. Ding, T. Zhu, Face image de-identification by feature space adversarial perturbation. Concurr. Comput. Practice Experience **35**(5), e7554 (2023)
10. Y. Wen, B. Liu, J. Cao, R. Xie, L. Song, Z. Li, IdentityMask: deep motion flow guided reversible face video de-identification. IEEE Trans. Circuits Syst. Video Technol. **32**(12), 8353–8367 (2022)
11. J. Cao, B. Liu, Y. Wen, R. Xie, L. Song, Achieving privacy-preserving multi-view consistency with advanced 3D-aware face de-identification, in *Proceedings of ACM Multimedia Asia* (ACM, New York, 2023), pp. 1–6

12. M. Shopon, A.S.M. Hossain Bari, Y. Bhatia, P.K. Narayanaswamy, S.N. Tumpa, B. Sieu, M. Gavrilova, Biometric system de-identification: concepts, applications, and open problems, in *Handbook of Artificial Intelligence in Healthcare: Vol 2: Practicalities and Prospects* (Springer, Berlin, 2022), pp. 393–422
13. A. Halder, P. Chattopadhzyay, S. Kumar, Gait transformation network for gait de-identification with pose preservation. Signal Image Video Process. **17**(5), 1753–1761 (2023)

Chapter 10
Conclusion

In the digital age, the proliferation of facial recognition technology has raised significant concerns about the privacy and security of individuals. Our book, "Face De-identification: Safeguarding Identities in the Digital Era," explores the critical topic of face de-identification techniques and their importance in preserving privacy and protecting identities.

Reflecting on our journey, we began by introducing the motivations behind face de-identification and the need to balance the pervasive use of facial recognition. We covered the fundamentals of face recognition and de-identification, setting the stage for subsequent discussions.

Part I delved into various face de-identification techniques, including obfuscation, k-Same, adversarial perturbation, and deep generative models. We also discussed evaluation metrics to measure privacy protection and utility preservation.

Part II featured in-depth analyses of specific methods and applications. We presented techniques like differential private k-anonymity, differential privacy for face images, personalized invertible de-identification, and high quality explainable models. Each method offered unique approaches to the challenge.

In our concluding Part III, we explored future prospects and challenges in face de-identification. The book touched on open research questions and technical challenges warranting further exploration. As technology evolves, so must our privacy preservation methods.

In summary, this book serves as a comprehensive guide on face de-identification for researchers, professionals, and policymakers. Our goal is to advance knowledge and contribute to the responsible use of facial data in an era where privacy and identity protection are paramount.

As challenges continue to evolve, protecting identities will remain fundamental. We encourage readers to stay vigilant, innovative, and dedicated to safeguarding identities in the digital era.

Y. Wen et al., *Face De-identification: Safeguarding Identities in the Digital Era*,
https://doi.org/10.1007/978-3-031-58222-6_10

Glossary

3D Morphable Model 3D Morphable Model (3DMM) is a statistical model used in computer vision and computer graphics to represent and analyze variations in facial shape and appearance within a population. It serves as a compact and versatile representation of facial geometry and texture, enabling the synthesis, analysis, and manipulation of facial shapes and textures.

Deep Generative Network A Deep Generative Network is a type of deep learning architecture that focuses on generating new data samples that are similar to those in the training dataset. These networks aim to model the underlying distribution of the training data and then generate synthetic data points that mimic the characteristics of the original dataset.

Deep Learning Deep learning is a subset of machine learning that utilizes artificial neural networks composed of multiple layers to extract and transform features from input data. It aims to model high-level abstractions in data using complex architectures consisting of many layers, allowing hierarchical representation learning.

Deep Neural Network A Deep Neural Network (DNN) is a type of artificial neural network (ANN) characterized by multiple layers between the input and output layers. It is designed to model complex patterns and relationships within data by utilizing a hierarchical or layered structure.

Differential Privacy Differential privacy is a framework and concept in data privacy and data analysis aimed at providing strong privacy guarantees for individuals while allowing useful information to be extracted from datasets. It ensures that the inclusion or exclusion of any single individual's data in a dataset will not significantly affect the outcome of queries or analyses.

Face De-identification Face de-identification, also known as face anonymization or face blurring, refers to the process of obscuring or modifying facial features in images or videos to protect the identity and privacy of individuals depicted in the visual content.

Y. Wen et al., *Face De-identification: Safeguarding Identities in the Digital Era*,
https://doi.org/10.1007/978-3-031-58222-6

Face Recognition Face recognition is a biometric technology that involves identifying or verifying individuals by analyzing and recognizing their facial features or patterns. It is a computer vision technology used in image analysis to automatically detect, locate, extract, and match facial characteristics from digital images or video frames.

Neural Radiance Field Neural Radiance Fields (NeRF) is a recent and innovative method in computer graphics and computer vision for synthesizing highly detailed and photo-realistic 3D scenes from 2D images or image collections. NeRF allows for the creation of immersive and realistic 3D representations of scenes by learning a continuous volumetric representation of the scene's geometry and appearance directly from 2D images.

Variational Autoencoder A Variational Autoencoder (VAE) is a type of generative model and a variant of the autoencoder neural network architecture. VAEs are designed for unsupervised learning and are capable of learning a latent representation of input data while simultaneously generating new data samples.

SPRINGER NATURE

GPSR Compliance

The European Union's (EU) General Product Safety Regulation (GPSR) is a set of rules that requires consumer products to be safe and our obligations to ensure this.

If you have any concerns about our products, you can contact us on ProductSafety@springernature.com

In case Publisher is established outside the EU, the EU authorized representative is:

Springer Nature Customer Service Center GmbH
Europaplatz 3
69115 Heidelberg, Germany

The manufacturer's authorised representative in the EU is Springer Nature Customer Service Centre GmbH, Europaplatz 3, 69115 Heidelberg, Germany. If you have any concerns regarding our products, please contact ProductSafety@springernature.com

Printed and bound by CPI Group (UK) Ltd, Croydon, CR0 4YY
30/09/2026
02225598-0002